NON-LINEAR WAVE MECHANICS

A CAUSAL INTERPRETATION

Original title:

UNE TENTATIVE D'INTERPRÉTATION CAUSALE ET NON LINÉAIRE DE LA MÉCANIQUE ONDULATOIRE

(LA THÉORIE DE LA DOUBLE SOLUTION)

Gauthier-Villars, Paris, 1956

NON-LINEAR WAVE MECHANICS

A CAUSAL INTERPRETATION

by

LOUIS DE BROGLIE

Member of the Académie Française
Permanent Secretary of the Académie des Sciences
Professor at the Sorbonne, Paris (France)

Translated by

ARTHUR J. KNODEL

Professor of French,
University of Southern California, Los Angeles, Calif. (U.S.A.)

and

JACK C. MILLER

Assistant Professor of Physics,
Pomona College, Claremont, Calif. (U.S.A.)

ELSEVIER PUBLISHING COMPANY

AMSTERDAM LONDON NEW YORK PRINCETON

1960

SOLE DISTRIBUTORS FOR THE UNITED STATES OF NORTH AMERICA:
D. VAN NOSTRAND COMPANY, INC.
120 Alexander Street, Princeton, N.J. (Principal Office)
257 Fourth Avenue, New York 10, N.Y.

SOLE DISTRIBUTORS FOR CANADA:
D. VAN NOSTRAND COMPANY (CANADA), LTD.
25 Hollinger Road, Toronto 16

SOLE DISTRIBUTORS FOR THE BRITISH COMMONWEALTH EXCLUDING CANADA:
D. VAN NOSTRAND COMPANY, LTD.
358 Kensington High Street, London, W. 14

With 20 figures

Library of Congress Catalog Card Number 59-12588

PRINTED IN THE NETHERLANDS BY
DIJKSTRA'S DRUKKERIJ N.V., VOORHEEN BOEKDRUKKERIJ GEBR. HOITSEMA, GRONINGEN

PREFACE

It is sometimes said that when a man grows old he goes back to the preoccupations of his youth. Perhaps that is why, for some four years now, I have asked myself the following question: Might not the concepts which guided my research from 1922 to 1928, when I first worked on Wave Mechanics, be more accurate and fundamental than those that have prevailed since that time?

As early as 1923 I had clearly seen that the propagation of a wave must be associated with the movement of every particle, but the continuous wave—of the type familiar in Classical Optics—which I had been led to consider and which became the Ψ wave of ordinary Wave Mechanics, did not seem to me to describe the physical reality accurately; only its *phase*, related directly to the motion of the particle, seemed to me of fundamental significance, and that is why I had named the wave which I associated with the particle "the phase-wave"—a designation that is completely forgotten today, but which at that time I believed entirely justified. However, as the work of other scientists led to further progress in Wave Mechanics, it became daily more evident that the Ψ wave with its continuous amplitude could be used only in statistical predictions. And so, little by little, there was an increasing trend towards the "purely probabilistic" interpretation, of which Born, Bohr and Heisenberg were the chief advocates. I was surprised at this development, which did not seem to me to fulfil the "explanatory" aim of theoretical physics; and that is what led me, around 1925—1927, to believe that all problems of Wave Mechanics required a set of two coupled solutions of the wave equation: one, the Ψ wave, definite in phase, but, because of the continuous character of its amplitude, having only a statistical and subjective meaning; the other, the u wave of the same phase as the Ψ wave but with an amplitude having very large values around a point in space and which, precisely on account of its spatial singularity (a singularity, moreover, which may not be one in the strict mathematical sense of the term) can be used to describe the particle objectively. In this way I obtained, in agreement with Einstein's concepts, what I had always believed must be sought: a picture of the particle in which it appears as the center of an extended wave phenomenon involving the particle in an intimate way. And, thanks to the theoretically postulated parallelism between u and Ψ waves, the Ψ wave, it seemed to me, preserved all the statistical properties that had quite rightly been attributed to it.

Such was the idea that had taken shape in my mind, and its curious subtlety astonishes me to this day. I called it the "theory of the Double Solution," and it was *that* idea which translated my real thinking in all its complexity. But, in order to facilitate explaining it, I had sometimes given it a simplified form, much less profound, to my way of thinking, which I had named "the pilot-wave theory", in which the particle, assumed given *a priori*, was considered to be piloted by the continuous Ψ wave. Discouraged by the rather unfavorable reception accorded my ideas by most theoretical physicists, who were strongly attracted by the formal elegance and apparent rigor of the purely probabilistic interpretation, I too came round to that interpretation and for more than twenty years admitted its correctness.

But, as I have said, since 1951 I have once again been wondering if, after all, my first idea was not the right one. Further reflections on this very difficult problem have led me to refine certain points of the original double-solution theory and, in certain other points actually to modify that theory, notably by introducing a hypothesis that today strikes me as indispensable: namely, that the equation of the propagation of the u wave is, basically, non-linear and, consequently, different from that admitted for the Ψ wave, even though the two equations may be considered identical *almost everywhere.*

In the present work, after a summary of the purely probabilistic interpretation, which is now "orthodox", and the objections to it put forward by a small but outstanding group of scientists, there will be found a general exposition of the present state of my reflections on the theory of the Double Solution. I take the liberty of particularly drawing the reader's attention to Chapters XVII, XVIII and XIX, which contain very bold suggestions, to be sure—but suggestions which may have very far-reaching consequences. I hope very much that young theorists gifted with physical insight, and experienced mathematicians as well, will be good enough to take an interest in the hypotheses which I have put forward in the conclusion of my work without being able to give real justification for them.

I have once more taken up the study of my former and earliest conceptions of Wave Mechanics without preconceived ideas of any sort and without any personal axe to grind. I may be wrong in wishing to go back to concepts that are clearer than those prevailing in theoretical physics at the present time. But I should like this line of thought, abandoned for some twenty-five years now and believed to lead to an impasse, to be carefully re-examined to see whether, on the contrary, it may not be the pathway that might lead to the true Microphysics of the Future.

Louis de Broglie

LOUIS DE BROGLIE

CONTENTS

PART ONE

The Basic Ideas and the Standard Purely Probabilistic Interpretation of Wave Mechanics

PART ONE

The Basic Ideas and the Standard Purely Probabilistic Interpretation of Wave Mechanics

Chapter I

THE BASIC IDEAS OF WAVE MECHANICS

1. Point of departure

The idea which, in my 1923-1924 works, served as the point of departure for Wave Mechanics was the following: Since for light there exists a corpuscular aspect and a wave aspect united by the relationship Energy = h times frequency, where h, Planck's constant, enters in, it is natural to suppose that, for matter as well, there exists a corpuscular *and* a wave aspect, the latter having been hitherto unrecognized. These two aspects must be united by the general formulas in which Planck's constant figures, and must contain as special cases those relationships applicable to light.

In order to elaborate this idea, it seemed to me in 1923 that it was necessary to associate a periodic element to the corpuscular concept. Let us imagine a particle whose motion is rectilinear and uniform in a certain direction in the absence of any external field. We will concentrate our attention exclusively on the state of motion, leaving aside its position in space. The motion will be in a certain direction, which we will take as the z-axis, and it will be defined by two quantities: energy and momentum, for which the relativistic expressions, as a function of the particle's rest-mass m_0, are given by the formulas

$$\mathrm{W} = \frac{m_0 c^2}{\sqrt{1-\beta^2}}, \qquad \mathbf{p} = \frac{m_0 \mathbf{v}}{\sqrt{1-\beta^2}} \qquad \left(\beta = \frac{v}{c}\right) \tag{1}$$

from which we deduce the relationship:

$$|\mathbf{p}| = p = \frac{\mathrm{W}}{c^2}|\mathbf{v}| = \frac{\mathrm{W}}{c^2}v. \tag{2}$$

The state of motion is thus defined for a certain observer A, fixed in a Galilean system of reference—an observer who employs a time t and rectangular coordinates x, y, z.

Let us now suppose another observer B who has, relative to the first observer, the speed v in the direction Oz—in other words, an observer moving with the particle. We may suppose that B has chosen an axis

$O_0 z_0$ moving along Oz, and axes $O_0 x_0$ and $O_0 y_0$ respectively parallel to Ox and Oy. This being the case, the space and time coördinates x_0, y_0, z_0 and t_0 of B are related to A's coördinates x, y, z and t by the well-known formulas of the simple Lorentz transformation:

$$x_0 = x, \; y_0 = y, \; z_0 = \frac{z - vt}{\sqrt{1-\beta^2}}, \; t_0 = \frac{t - \frac{\beta}{c} z}{\sqrt{1-\beta^2}}. \tag{3}$$

Now, for observer B, the speed of the particle is zero: this gives the values for energy and momentum

$$W = m_0 c^2; \qquad \mathbf{p} = 0 \tag{4}$$

Pursuing our basic idea, we should now seek to introduce a periodic element, and we will attempt to define it first in the particle's own system, that is, in the system of reference of observer B. Since everything is at rest in this system, it is natural to define any desired periodic element in it in the form of a stationary wave. To do that, we will define the periodic element by the assumed scalar magnitude

$$\Psi_0 = a_0 e^{2\pi i \nu_0 t} \tag{5}$$

which has the form of the complex representation of a stationary wave. Ψ_0 oscillates as a function of the time t_0 with a frequency ν_0, characteristic of the nature of the particle under consideration. Let us suppose that a_0 is a constant (in general, complex) so that Ψ_0 has the same value for every point of observer B's system at the instant t_0.

We can represent the distribution of the values of Ψ_0 by imagining an infinite number of tiny clocks placed at every point in the particle's system, synchronized with respect to each other, and having a period $T_0 = 1/\nu_0$. These tiny clocks represent, so to speak, at every point the "phase" of the periodic phenomenon—a phase which is the same everywhere for observer B at a single instant t_0 of his own time.

What value should we give to the frequency ν_0? We should obviously seek to define it in terms of a magnitude that characterizes the particle in the B-system. Now, in that system we have only one magnitude that is not of zero value, the energy $W_0 = m_0 c^2$. In view of the role played by Planck's constant h in all quantum problems, it is natural to postulate

$$\nu_0 = \frac{W}{h} = \frac{m_0 c^2}{h} \tag{6}$$

analogous to the Einstein relationship for photons.

How will the periodic element that we have just defined for observer B appear to observer A? Assuming, as is natural here, that the element Ψ is an invairant, all one has to do in order to obtain its expression for A is to substitute in its B-form the value t_0 furnished by the fourth Lorentz equation—equation (3) here—giving us

$$\Psi(x, y, z, t) = a_0 e^{2\pi i\nu\left(t-\frac{z}{V}\right)} \tag{7}$$

if we put

$$\nu = \frac{\nu_0}{\sqrt{1-\beta^2}}, \qquad V = \frac{c}{\beta} = \frac{c^2}{v}. \tag{8}$$

Thus, for observer A, who sees the particle go by at the speed v in the direction Oz, the phases of the periodic Ψ phenomenon are distributed like those of a plane monochromatic wave whose frequency ν and phase velocity V would have the values indicated in (8).

This can be expressed in another way by going back to the picture of an infinite number of tiny clocks distributed at every point in space and having the same phase for observer B. As a consequence of the relativistic phenomenon of the slowing down of clocks in motion, each one of these clocks appears to observer A as having a *diminished* frequency

$$\nu_A = \nu_0\sqrt{1-\beta^2} \tag{9}$$

but the distribution in the phases of all the clocks is given for A by formula (7), which means that it coincides with the distribution of the phases of a plane monochromatic wave of frequency ν and phase velocity V given by (8).

Comparing formulas (8) and (9), one notes the essential difference between the apparent frequency ν_A of a clock in motion—a frequency which is *diminished* on account of the motion—and the frequency ν of the wave associated with this motion which is *increased* by that motion.

This difference between the relativistic variations of the frequency of a clock and the frequency of a wave is fundamental; it had greatly attracted my attention, and thinking over this difference determined the whole trend of my research.

The foregoing may be summed up by saying that the particle identified with one of the tiny clocks moves, in relation to the phase of the wave, with the velocity $V - v = c(1 - \beta^2)/\beta$ in such a way as to remain constantly in phase with the wave.

Let us consider this last idea in a more precise form. Among the infinite number of tiny clocks that we have imagined, let us suppose there is one that plays a special part. That one is the regulator-clock which we will identify with the particle, while the other clocks represent the phases of the wave phenomenon of which the particle would be the center. Within their own system, all the clocks are at rest and have the same frequency ν_0. In the system of the observer who sees all the clocks go by at a velocity v, the totality of the phases of these clocks is given by the factor $\nu(t-(z/V))$ with the definitions of (8). During a time dt, the regulator-clock moves a distance $v\,dt$ in the direction Oz and its variation proportionally is $\nu_0\sqrt{1-\beta^2}\,dt$. The phase of the wave at the point where the clock is situated varies by $\nu_0/\sqrt{1-\beta^2}(dt-(v\,dt/V))$. Since these two variations must be equal, we must have

$$\sqrt{(1-\beta^2)} = \frac{1}{\sqrt{1-\beta^2}}\left(1-\frac{v}{V}\right) \quad \text{or } \beta^2 = \frac{v}{V} \tag{10}$$

in agreement with the second relationship (8).

But let us put aside these pictures, which we will return to later on, and let us consider once more the formulas obtained. A comparison of the first relationships (1) and (8) gives us

$$W = h\nu \tag{11}$$

a relationship that must obviously hold for the whole Galilean system, since observer A is an arbitrary Galilean observer.

Defining the wave-length of the Ψ wave in the usual way by the formula $\lambda = V/\nu$, we obtain the value:

$$\lambda = \frac{c^2}{v}\frac{h}{W} = \frac{h}{p} \tag{12}$$

We have thus found the two basic formulas (11) and (12) which define the frequency and the wave-length of the wave associated with the particle, in terms of the energy and the momentum of that particle. For very low velocities, as compared with the velocity of light in vacuum, formula (12) takes on the approximate form:

$$\lambda = \frac{h}{mv}. \tag{13}$$

For a particle of velocity equal to c (or indistinguishable from c), we have

$$v = V = c, \qquad W = h\nu, \qquad p = \frac{h\nu}{c} = \frac{h}{\lambda} \tag{14}$$

That, in fact, is the way the basic formulas of the theory of light-quanta (Einstein, 1905) are found applicable to photons.

We can now write the magnitude Ψ, as measured by observer A, in the form

$$\Psi = a_0 e^{\frac{i}{\hbar}(Wt - pz)} \qquad \hbar = \frac{h}{2\pi} \tag{15}$$

and, more generally, if one has not taken the direction of propagation as the z-axis

$$\Psi(x, y, z, t) = a_0 e^{\frac{i}{\hbar}(Wt - p_x x - p_y y - p_z z)} = a_0 e^{\frac{i}{\hbar}(Wt - \mathbf{p} \cdot \mathbf{r})} \tag{16}$$

a formula which shows that the phase of the Ψ wave is, apart from the factor $2\pi/h$, equal to the Hamiltonian action of the particle. Noting this proportional relationship between the action of the particle and the phase of the Ψ wave associated with it, we perceive that the principle of stationary action for the Dynamics of the particle must be an equivalent of Fermat's principle that will be valid for the associated wave. But wave-theory teaches us that Fermat's principle is valid only in the area where geometrical Optics may be used and that it has no meaning in the area of exclusively undulatory physical Optics. As early as 1923 I had formulated the fundamental idea that traditional Mechanics (in both its relativistic and classically Newtonian form) was only an approximation with the same range of validity as geometrical Optics. From that moment I was led to realize the necessity of constructing a new Mechanics, an undulatory Mechanics "which would be to traditional Mechanics what undulatory Optics is to geometrical Optics". Such was the starting point of Wave Mechanics.

2. First developments in Wave Mechanics

At the time when the ideas which I have just summarized occurred to me, I was imbued with classical ideas of the possibility of representing phenomena in an objective and deterministic way within the framework of space-time. It therefore seemed to me that the wave-particle association must take the following form: the particle would be some sort of singularity at the center of an extended wave phenomenon—a phenomenon of which it would be an integral part, and the motion of this singularity, although almost certainly behaving according to the new dynamical laws, must involve, in conformity with the classical picture, a trajectory in space and a velocity determined for every point of this trajectory.

So it occurred to me that the plane monochromatic Ψ wave, associated in my earlier reasoning with the uniform and rectilinear motion of a free particle, did not actually describe reality, but that it could give in a precise way only the *phase* of the wave phenomenon surrounding the particle, since the constant amplitude a_0 could not represent the true amplitude of this phenomenon. Indeed, the true amplitude seemed to me to involve a singularity—the particle; and this amplitude had to diminish with the distance from that singularity. The true wave-function representing the combination of the undulatory phenomenon and its singularity seemed to me necessarily to be, in the case of rectilinear and uniform motion, of the form:

$$u(x, y, z, t) = f(x, y, z, t)e^{\frac{i}{\hbar}\varphi(x, y, z, t)} \tag{17}$$

φ being the phase $Wt - \mathbf{p} \cdot \mathbf{r}$ and $f(x, y, z, t)$ a function involving a singularity moving with the velocity v. The wave function

$$\Psi = a_0 e^{\frac{i}{\hbar}\varphi(x, y, z, t)} \tag{18}$$

would have a true phase φ which would be that of our tiny imaginary clocks carried along by the movement of the particle, but its constant amplitude would not describe the real distribution of the wave phenomenon in space. At most it may represent a kind of statistical average when one has no way of knowing which of the straight lines parallel to the direction of the movement is really described by the particle and at what point of the trajectory it is at the time t.

It is because the phase φ seemed to me to have a profound physical meaning connected with relativistic effects, and because I felt that this phase should be found in the real wave-function u, that I called the Ψ function the "phase-wave", hoping thereby to reserve for a later date a more thorough investigation of the question of the meaning of its amplitude.

The ideas that I have just reviewed were adopted by me in all my early expositions of Wave Mechanics from 1924 to 1927. They were to lead me in 1927 to the theory of the double solution, to which the second half of the present Work is devoted. In the meantime, however,

Schrödinger, in a series of admirable studies, had succeeded in making considerable progress in the mathematical formalism of the new Wave Mechanics. He had written out its general equations and had applied them to the calculation of stationary states of quantized systems, thus replacing by a rigorous theory the intuitive justification, which I had given in my early works, of the formulas of quantization in the old quantum theory. Finally, he had shown the fundamental identity of the methods of Wave Mechanics and the matrix formulations of Quantum Mechanics developed in 1927 by Heisenberg—an identity that had been hidden by the difference in their mathematical aspects.

In order to make such studies in Wave Mechanics, Schrödinger in particular utilized the analogy between Analytic Mechanics and Geometrical Optics, which had been pointed out by Hamilton a century earlier, all the while taking into account, of course, the existence of quanta and using as his point of departure the ideas which I had put forward in my early works and summarized in my Doctoral Dissertation of 1924.

I will now give an exposition of this second method of approaching Wave Mechanics—a path that Schrödinger followed, limiting himself to the Newtonian approximation and not taking the Relativistic corrections into account.

Chapter II

THE HAMILTONIAN APPROACH TO WAVE MECHANICS (THE ANALOGY BETWEEN ANALYTICAL MECHANICS AND GEOMETRICAL OPTICS)

1. Classical mechanics of the point mass. Jacobi's theorem

Following the older conceptions, a particle (point mass) must have at every instant a well-determined position in space which develops in time; under the influence of a field of force to which it is subjected, it describes a certain curve in space, which is its trajectory. Moreover, we here assume that the field of force is derivable from a potential $V(x, y, z, t)$. At every instant the position of the particle in space is located by three spatial coördinates $x(t)$, $y(t)$, $z(t)$. The classical equations of motion are then as follows:

$$m\frac{d^2x}{dt^2} = -\frac{\partial V}{\partial x}, \quad m\frac{d^2y}{dt^2} = -\frac{\partial V}{\partial y}, \quad m\frac{d^2z}{dt^2} = -\frac{\partial V}{\partial z} \tag{1}$$

m being a constant characteristic of the particle and called its mass.

The three differential equations (1) of the second order in t define completely the change in the coördinates with time, that is, its motion, when *six* arbitrary constants are assumed, representing the coördinates and the components of velocity at a given instant t_0, referred to as the "initial instant". The determinism of the older Mechanics resides in the fact that, assuming the initial state of position and velocity to be known, all subsequent states are rigorously determined.

We refer the reader to the treatises on Rational Mechanics for the proofs of the general theorems of the Dynamics of a point mass and for the theory of the equations of Lagrange, Hamilton, etc. We will here confine ourselves to a statement of Jacobi's theorem, which will be useful in subsequent developments:

THEOREM. — *If one succeeds in finding a complete integral (that is, a solution depending on three non-additive arbitrary constants α, β, γ) for the partial differential equation (Jacobi's equation)*

$$\frac{1}{2m}\left[\left(\frac{\partial S}{\partial x}\right)^2 + \left(\frac{\partial S}{\partial y}\right)^2 + \left(\frac{\partial S}{\partial z}\right)^2\right] + V(x, y, z, t) = \frac{\partial S}{\partial t} \tag{2}$$

the equations

$$\frac{\partial S}{\partial \alpha} = a, \quad \frac{\partial S}{\partial \beta} = b, \quad \frac{\partial S}{\partial \gamma} = c; \tag{3}$$

where a, b, c are new arbitrary constants, define one of the possible motions of the particle in the field of force. The components of momentum of the particle when in such motion at time t and position x, y, z, are given by the relations

$$p_x = mv_x = -\frac{\partial S}{\partial x}, \quad p_y = mv_y = -\frac{\partial S}{\partial y}, \quad p_z = mv_z = -\frac{\partial S}{\partial z} \tag{4}$$

that is

$$\mathbf{p} = m\mathbf{v} = -\operatorname{grad} S.$$

Hence we see that, according to Jacobi's theorem, the possible motions of the particle are divided into classes, the motions of any one class corresponding to one complete integral of Jacobi's equation $S(x, y, z, t, \alpha, \beta, \gamma)$ with values given for the "primary constants" α, β, γ. Each of these classes contains an infinity of possible motions, each of which is characterized by the value of the "secondary constants" a, b, c.

We must remember that Jacobi's equation may be obtained by starting with the energy as a function of the coördinates and the conjugate momenta (the Hamiltonian expression of energy).

$$E(x, y, z, p_x, p_y, p_z, t) = \frac{1}{2m}(p_x^2 + p_y^2 + p_z^2) + V(x, y, z, t) \tag{5}$$

and, by replacing p_x, p_y, p_z respectively by $-\partial S/\partial x$, $-\partial S/\partial y$, $-\partial S/\partial z$, and then by setting the expression thus obtained equal to $\partial S/\partial t$.

Jacobi's theorem takes on a particularly simple form in the important case where the potential function V does not depend explicitly on time. We know that in this case there is conservation of energy, that is to say that during the course of the motion the sum of the kinetic and potential energy, $\frac{1}{2}mv^2 + V$, has a constant value E. The constant E here plays the role of one of the primary constants, for example γ. If we then put

$$S(x, y, z, t, \alpha, \beta, E) = Et - S_1(x, y, z, \alpha, \beta, E) \tag{6}$$

where S_1 no longer depends on time, we will have to seek a complete integral dependent on the constant E and on two other constants α and β in the partial differential equation (called the "separated"

Jacobi's equation).

$$\frac{1}{2m}\left[\left(\frac{\partial S_1}{\partial x}\right)^2+\left(\frac{\partial S_1}{\partial y}\right)^2+\left(\frac{\partial S_1}{\partial z}\right)^2\right]+V(x, y, z)=E \qquad (7)$$

The general theorem of Jacobi applied to this special case shows us that, if we have found a complete integral solution of equation (7), the motion defined by the equations

$$\frac{\partial S_1}{\partial \alpha}=a, \quad \frac{\partial S_1}{\partial \beta}=b, \quad \frac{\partial S}{\partial E}=t-\frac{\partial S_1}{\partial E}=c \qquad (8)$$

where a, b, c are three arbitrary constants, will be one of the possible motions of the particle in a static field of force. As for the momentum of the particle at the time of its passage through the point x, y, z, it will be given by

$$\mathbf{p}=m\mathbf{v}=\text{grad } S_1 \qquad (9)$$

The possible motions are thus sorted into classes, each of which corresponds to a single value of the energy and of the two primary constants α and β; each class comprises an infinity of possible motions, each of which is characterized by the value of the three secondary constants a, b, c.

The first two equations of (8) do not involve time; they define a curve in space which is the trajectory of the particle. The third equation of (8), which may be written $\partial S_1/\partial E = t - t_0$, gives the motion along the trajectory (equation of time). Thus one sees that in the case of static fields, the trajectory may be studied independent of the motion — a situation that does not hold in the case of fields that vary with time.

Another important valid result in the case of static fields is the following: *The trajectories of a single class which correspond to the same complete integral* $S_1(x, y, z, \alpha, \beta, E)$ *are orthogonal to the surfaces* $S_1 =$ *constant.* This follows directly from equation (9).

This property of the trajectories of being normal to the surfaces $S_1 =$ const. leads us back to Maupertuis' principle of least action. To see this, let us consider all the surfaces $S_1 =$ const., with the constant lying between C_1 and C_2 which are separated infinitesimally from each other, and let us represent a few of these surfaces in cross-section.

Let AEB be a trajectory of the class corresponding to S_1 and AFB a curve infinitesimally separated from AEB. If we use dn to designate the differential displacement normal to the surfaces $S_1 =$ const., the integral $\int (\partial S_1/\partial n)ds$ taken along AEB is equal to $C_2 - C_1$, the element of the curve ds here being equal to dn. Let us take the same integral along AFB. The contribution to this integral of a small element such as FG is greater or at least equal to the change in S_1 from F to G. In

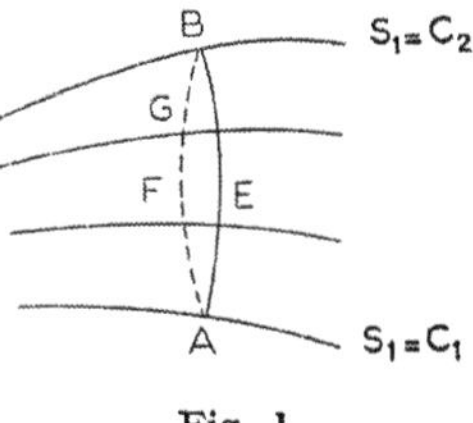

Fig. 1

fact, if $\overline{FG}$ is normal to the surfaces $S_1 = \text{const.}$ which pass through its extremities, one has $\overline{FG} = dn$ and $(\partial S_1/\partial n)\overline{FG} = S_1(G) - S_1(F)$, whereas if $\overline{FG}$ is not normal to the surfaces $S_1 = \text{const.}$, one has $\overline{FG} > dn$, and $(\partial S_1/dn)\overline{FG}$ is greater than $S_1(G) - S_1(F)$. Now all the elements of AFB cannot be normal to the surfaces $S_1 = \text{const.}$, otherwise AFB coincides with the trajectory AEB. Thus the integral $\int_A^B (\partial S_1/\partial n)ds$ is greater along AFB than along AEB.

According to equation (7), we have

$$\frac{\partial S_1}{\partial n} = \left[\left(\frac{\partial S_1}{\partial x}\right)^2 + \left(\frac{\partial S_1}{\partial y}\right)^2 + \left(\frac{\partial S_1}{\partial z}\right)^2\right]^{\frac{1}{2}} = \sqrt{2m(E - V(x, y, z))} \qquad (10)$$

We thus arrive at the following statement: *The trajectory passing through two points* A *and* B *in space is characterized by the fact that the integral* $\int_A^B ds\sqrt{2m(E - V)}$ *is smaller for the real trajectory than for every curve infinitesimally close to it going from* A *to* B. This is Maupertuis' principle of least action.[1]

It is possible to illustrate the foregoing considerations by a very simple example. Let us consider the motion of a particle in the absence of any field. Then $V = 0$ and, since there is conservation of energy, we may write equation (7) in the form

$$\frac{1}{2m}\left[\left(\frac{\partial S_1}{\partial x}\right)^2 + \left(\frac{\partial S_1}{\partial y}\right)^2 + \left(\frac{\partial S_1}{\partial z}\right)^2\right] = E. \qquad (11)$$

[1] Note that the curve AFB must be continuous, as well as its slope. Note also that the reasoning is invalid when the trajectories have an envelope and when AFB touches that envelope between A and B. Maupertuis' integral is always stationary in that case, but it may be maximum instead of minimum.

We obtain a complete integral if we put

$$S_1 = \sqrt{2mE}(\alpha x + \beta y + \gamma z) \text{ with } \gamma^2 + \beta^2 + \alpha^2 = 1 \quad (12)$$

According to Jacobi's theorem, we obtain the trajectories by writing

$$\frac{\partial S_1}{\partial \alpha} = \sqrt{2mE}\left(x - \frac{\alpha}{\gamma}z\right) = a, \quad \frac{\partial S_1}{\partial \beta} = \sqrt{2mE}\left(y - \frac{\beta}{\gamma}z\right) = b \quad (13)$$

They are, thus, straight lines with direction cosines α, β, γ, normals to the surfaces $S_1 =$ const. The motion along one of these straight lines is defined by the equation for the time

$$\frac{\partial S_1}{\partial E} = \frac{m}{\sqrt{2mE}}(\alpha x + \beta y + \gamma z) = t - t_0 \quad (14)$$

The motion is thus rectilinear uniform and takes place with velocity $v = \sqrt{2E/m}$.

Finally, the relations $p_x = mv_x = m\alpha v = \alpha\sqrt{2mE} = \partial S_1/\partial x, \ldots$ are easily verifiable. The desired complete integral thus defines the class of rectilinear uniform motions in the direction α, β, γ and of velocity $\sqrt{2E/m}$.

2. Propagation of waves in an isotropic medium. The approximation of Geometrical Optics

In order to facilitate the transition from Classical Mechanics to Wave Mechanics, let us make a rapid study of the propagation of monochromatic waves in a medium that is refractive, isotropic and dispersive, with properties that are constant in time. We will assume that this propagation is governed by the equation

$$\Delta\Psi = \frac{1}{\mathscr{V}^2}\frac{\partial^2\Psi}{\partial t^2} \quad (15)$$

Ψ being the wave-function and $\mathscr{V}$ a magnitude that is generally a function of the point x, y, z and of the frequency ν of the wave. $\mathscr{V}$ is the velocity of the propagation of the phase or, more briefly, the "velocity of propagation". For a monochromatic wave, that is, one with a well determined frequency ν, we will write

$$\Psi(x, y, z, t) = a(x, y, z)e^{2\pi i\nu t} \quad (16)$$

and to define the index of refraction n of the medium we will put

$$\frac{1}{\mathcal{V}} = \frac{n(x, y, z, \nu)}{\mathcal{V}_0} \tag{17}$$

$\mathcal{V}_0$ being the velocity of propagation in a reference medium where the index of refraction can be taken to be one. We then have

$$\Delta\Psi + \frac{4\pi^2 n^2 \nu^2}{\mathcal{V}_0^2}\Psi = 0 \tag{18}$$

Strictly speaking, the study of the propagation of a monochromatic wave in a dispersive medium, having permanent properties, should be made by determining the solution of equation (18) which corresponds to the required boundary conditions, but it often happens that one can solve the problem by an approximating procedure which is the basis of Geometrical Optics.

In order to understand fully the meaning of this approximation, let us first consider the case where the index n does not depend on x, y, z (a homogeneous medium). One then obtains a rigorous solution of equation (18) by putting

$$\Psi = ae^{2\pi i\nu\left[t - \frac{n(\nu)}{\mathcal{V}_0}(\alpha x + \beta y + z\sqrt{1-\alpha^2-\beta^2})\right]} \tag{19}$$

a being a constant called the amplitude of the wave. This solution (19) represents a "plane monochromatic wave". We shall call the linear function

$$\varphi(x, y, z, t) = \nu t - \frac{n\nu}{\mathcal{V}_0}(\alpha x + \beta y + z\sqrt{1 - \alpha^2 - \beta^2}) = \nu t - \varphi_1(x, y, z) \tag{20}$$

the phase of this wave.

The surfaces of equal phase $\varphi =$ const., also called "wave surfaces", are planes normal to the direction of the direction cosines α, β, $\gamma = \sqrt{1 - \alpha^2 - \beta^2}$. In the course of time the values of the phase, and consequently of the Ψ function, will move in that direction with the velocity

$$\mathcal{V} = \frac{\mathcal{V}_0}{n(\nu)} \tag{21}$$

At a given instant, one will find the same value for Ψ on the planes of equal phase which are separated from one another by the distance

$$\lambda = \frac{\mathcal{V}_0}{n\nu} = \frac{\mathcal{V}}{\nu} \tag{22}$$

called "wave-length" and, at a given point, one will find the same values of Ψ at time-intervals equal to the period $T = 1/\nu$.

Let us now consider a non-homogeneous medium where the index varies with x, y, z, and yet varies slowly enough to remain approximately constant *over a distance comparable to a wave-length*. It is then easy to see that the derivatives of $a(x, y, z)$ will be negligible in comparison with those of $\varphi_1(x, y, z)$ if one has the monochromatic wave (with real a)

$$\Psi(x, y, z, t) = a(x, y, z)e^{2\pi i[\nu t - \varphi_1(x, y, z)]} \tag{23}$$

and, by substituting (23) into (18), one will obtain an approximate equation that is known by the name of "the equation of Geometrical Optics":

$$\left(\frac{\partial\varphi_1}{\partial x}\right)^2 + \left(\frac{\partial\varphi_1}{\partial y}\right)^2 + \left(\frac{\partial\varphi_1}{\partial z}\right)^2 = \frac{\nu^2 n^2(x, y, z)}{\mathcal{V}_0^2} \tag{24}$$

Equation (24) permits us to determine the variations of the phase of the monochromatic wave (which is now no longer assumed linear in x, y, z) without having to take into account the variations of the amplitude $a(x, y, z)$ which are small over a distance comparable to a wave-length.

Now let $\varphi_1(x, y, z, \alpha, \beta, \nu)$ be a complete integral of equation (24), dependent on the three constants ν, α, β. The function,

$$\Psi = a e^{2\pi i[\nu t - \varphi_1(x, y, z, \nu, \alpha, \beta)]}$$

where a is slowly varying over a distance comparable to a wave-length then represents an approximate solution of the equation of propagation. By definition, the curves orthogonal to the surfaces $\varphi_1 =$ const. are the "rays" of the wave. In the same way that Maupertuis' principle of least action was previously justified for the particle-trajectories normal to the surfaces $S_1 =$ const., we can here demonstrate in like fashion "Fermat's principle", according to which, if a curve C is a ray of a wave-propagation passing through two points A and B in space, the integral

$$\int_A^B ds \frac{\partial\varphi_1}{\partial n} = \int_A^B ds \frac{n\nu}{\mathcal{V}_0}$$

taken along the ray C is smaller than the same integral taken along a curve infinitesimally close to C and joining the points A and B.[2]

[2] In order to avoid all possible confusion, it must be clearly understood that $\partial\varphi_1/\partial n$ is the derivative of φ_1 along the normal to the surface $\varphi_1 =$ const (and not the derivative of φ_1 in relation to the index n).

Geometrical Optics is merely an approximation, valid only if the index n changes very slightly over a distance comparable to the local wave-length defined by $\lambda = (\partial\varphi_1/\partial n)^{-1}$. When the wave-length approaches zero, this approximation tends towards rigorous accuracy.

We must now turn our attention to the presence of the frequency ν in equation (18). Instead of considering a monochromatic wave, we may consider, since the equation of propagation (15) is linear in Ψ, a sum or, as we say, a "superposition" of monochromatic waves, each obeying equation (18). It is thus desirable to find a form of the equation of propagation in which frequency does not enter and which is satisfied by the wave-function, even when it is made up of a superposition of monochromatic waves.

To give an example, let us assume that the index is given by a law of dispersion

$$n(x, y, z, \nu) = \sqrt{1 - \frac{F(x, y, z)\mathscr{V}_0^2}{4\pi^2\nu^2}} \tag{25}$$

where F is a certain function of position. Then, one may take as the equation of propagation which is independent of ν,

$$\Delta\Psi - \frac{1}{\mathscr{V}_0^2}\frac{\partial^2\Psi}{\partial t^2} = F(x, y, z)\Psi \tag{26}$$

since, for a monochromatic wave of the form (16), equation (18) again results. We will find equations analogous to (26) in Wave Mechanics.

3. The transition from Classical to Wave Mechanics

We pointed out in the first two paragraphs of this chapter a great analogy in form between the Analytic Dynamics of the point mass and Geometrical Optics. The analogy was noticed more than a century ago by Hamilton, before Jacobi had stated it in precise terms. Today it may lead us to an understanding of the synthesis achieved by Wave Mechanics.

To that end, let us begin by comparing the motion of a particle in the absence of any field (V = 0) with the propagation of a wave in a homogeneous medium where the index n is independent of x, y, z. For the particle in the absence of any field we have found the separated Jacobi function

$$S_1 = \sqrt{2mE}(\alpha x + \beta y + \gamma z) = mv(\alpha x + \beta y + \gamma z); \quad (\alpha^2 + \beta^2 + \gamma^2 = 1). \tag{27}$$

On the other hand, for a monochromatic wave in a homogeneous medium, since the wave-length λ is then constant, one can write the phase φ_1, in terms of the approximation of Geometrical Optics, in the form

$$\varphi_1 = \frac{1}{\lambda}(\alpha x + \beta y + \gamma z) \tag{28}$$

α, β, γ being the same as in (27), if we assume that the direction of the particle's motion coincides with that of the propagation of the wave. The complete S and φ functions are then

$$\mathrm{S} = \mathrm{E}t - mv(\alpha x + \beta y + \gamma z), \quad \varphi = \nu t - \frac{1}{\lambda}(\alpha x + \beta y + \gamma z). \tag{29}$$

It is in the spirit of quantum theory to put $\mathrm{E} = h\nu$, that is, to associate the propagation of a wave of frequency $\nu = \mathrm{E}/h$ with the motion of a particle having energy E. This leads us to put

$$\varphi = \frac{\mathrm{S}}{h} \tag{30}$$

If, by hypothesis, we admit this relation, we obtain the two fundamental formulas of Wave Mechanics (non-relativistic)

$$\mathrm{E} = h\nu, \qquad \lambda = \frac{h}{mv} \tag{31}$$

In other words, with the rectilinear and uniform motion of the particle with energy E and momentum mv, we associate the propagation, in the direction of this motion of a plane monochromatic wave having the frequency E/h and a wave-length h/mv—a wave expressed by

$$\Psi = a e^{\frac{i}{\hbar}\mathrm{S}} \qquad (a \text{ constant}) \tag{32}$$

with S given by (29).

This correspondence between wave and motion becomes general in the case of the motion of a particle in a *static* field defined by the potential function $\mathrm{V}(x, y, z)$. So it becomes necessary to compare this motion with the propagation of a wave in a non-homogeneous medium where the index of refraction n and, consequently, the wave-length λ vary from one point to another. The expressions to be compared in the Jacobi function and the phases are

$$\mathrm{S} = \mathrm{E}t - \mathrm{S}_1(x, y, z) \qquad \varphi = \nu t - \varphi_1(x, y, z) \tag{33}$$

the functions S_1 and φ_1 being, respectively, complete integrals of the equations

$$\begin{aligned} &\left(\frac{\partial \mathrm{S}_1}{\partial x}\right)^2 + \left(\frac{\partial \mathrm{S}_1}{\partial y}\right)^2 + \left(\frac{\partial \mathrm{S}_1}{\partial z}\right)^2 = 2m[\mathrm{E} - \mathrm{V}(x, y, z)]; \\ &\left(\frac{\partial \varphi_1}{\partial x}\right)^2 + \left(\frac{\partial \varphi_1}{\partial y}\right)^2 + \left(\frac{\partial \varphi_1}{\partial z}\right)^2 = \frac{1}{\lambda^2(x, y, z)} \end{aligned} \tag{34}$$

It is entirely natural to again make the hypothesis here which is expressed by (30) and consequently to put

$$\mathrm{E} = h\nu, \qquad \mathrm{S}_1 = h\varphi_1. \tag{35}$$

The second formula easily gives us

$$\lambda = \frac{1}{|\mathrm{grad}\ \varphi_1|} = \frac{h}{|\mathrm{grad}\ \mathrm{S}_1|} = \frac{h}{\sqrt{2m(\mathrm{E} - \mathrm{V}(x, y, z))}} \tag{36}$$

and, since at every point one must have $\mathrm{E} = \frac{1}{2}mv^2 + \mathrm{V}(x, y, z)$, we again end up with the second equation (31), but here v and λ vary from one point to another.

How shall we write the equation of propagation which corresponds to the motion in a static field? Let us write equation (18) in the form

$$\Delta\Psi + \frac{4\pi^2}{\lambda^2(x, y, z)}\Psi = 0 \tag{37}$$

and substitute for λ the expression given by (36); we have

$$\Delta\Psi + \frac{2m}{\hbar^2}[\mathrm{E} - \mathrm{V}(x, y, z)]\Psi = 0 \tag{38}$$

By putting $\mathrm{V} = 0$, we once more have the equation valid when there is no field.

In every case where Geometrical Optics is sufficient to describe the propagation of the Ψ wave, we may write

$$\Psi = ae^{\frac{i}{\hbar}\mathrm{S}} = ae^{\frac{i}{\hbar}[\mathrm{E}t - \mathrm{S}_1(x, y, z)]} \tag{39}$$

and the trajectories, determined by the Classical Dynamics of a point mass, normal to the surfaces $\mathrm{S}_1 = \text{const.}$, will be simply the rays of propagation of the Ψ wave, normal to the surfaces $\varphi_1 = \text{const.}$

And thus we arrive at one of the main ideas of the new Mechanics. Whereas the older Mechanics attributed a rigorous character to these equations and considered them valid everywhere, the new Mechanics gives the leading role to the wave. It considers the equations of the older Mechanics as approximations valid only when the approximation of Geometrical Optics is sufficient to describe the propagation of the wave.

Classical Dynamics thus appears to be only an approximation. It is applicable only when the index n relative to the Ψ wave varies only slightly in comparison with a wave-length or, what is essentially the same thing, when the potential V varies slowly over the distance of a wave-length. However, if the wave-length of the Ψ wave were infinitesimally small, the older Dynamics would be rigorously valid. According to formula (37), we see that this condition would always be satisfied for a non-zero velocity v, if h were infinitesimally small. Whence the following long-familiar conclusion: If h is made to approach zero in the formulas, all quantum effects must disappear and Classical Dynamics again takes on its full rigor.

4. General equations for the Wave Mechanics of a particle

We have just been led to substitute for the classical equations of the Dynamics of a point mass in a constant field, the equation of the propagation of a monochromatic wave. But, as we will presently see, we are led to consider wave-trains formed by a superposition of monochromatic waves. It is therefore useful to try and obtain an equation of propagation which the Ψ function satisfies when it represents such a superposition of monochromatic waves. The equation

$$\Delta\Psi - \frac{2m}{\hbar^2}V(x, y, z)\Psi = i\frac{2m}{\hbar}\frac{\partial\Psi}{\partial t} \qquad (40)$$

satisfies this condition, because, for a plane monochromatic wave of frequency E/h, it brings us back to equation (38). But the form of (40) allows us to go beyond single monochromatic waves and to consider superpositions of such waves. In addition, it suggests the way to extend the new Mechanics to the case of fields varying with time. Indeed, since it permits us to go beyond monochromatic waves, time no longer plays a special part, and it is then natural to admit that the form of the equation must be preserved when V depends on time, and thus to write

$$\Delta\Psi - \frac{2m}{\hbar^2}V(x, y, z, t)\Psi = i\frac{2m}{\hbar}\frac{\partial\Psi}{\partial t} \qquad (41)$$

as the general form of the equation of propagation of Ψ waves in the non-relativistic Wave Mechanics of a single particle.

5. An automatic process for arriving at the wave-equation

We now indicate a formal process which permits us to arrive automatically at the wave-equation—a process whose importance was emphasized by Schrödinger.

In Classical Mechanics we call the function that expresses energy by means of coördinates and conjugate Lagrange momenta the "Hamiltonian function." In rectangular coördinates, the well-known expression of this function is

$$H(x, y, z, p_x, p_y, p_z, t) = \frac{1}{2m}(p_x^2 + p_y^2 + p_z^2) + V(x, y, z, t) \tag{42}$$

If in this expression we replace p_x by $-(\hbar/i)\partial/\partial x$, p_y by $-(\hbar/i)\partial/\partial y$ and p_z by $-(\hbar/i)\partial/\partial z$, we obtain an operator, the Hamiltonian operator

$$H_{op} = -\frac{\hbar^2}{2m}\left(\frac{\partial^2}{\partial x^2} + \frac{\partial^2}{\partial y^2} + \frac{\partial^2}{\partial z^2}\right) + V(x, y, z, t) \tag{43}$$

By applying operator (43) to the Ψ function, that is, by multiplying Ψ *in front* by the operator (43), and then setting the result equal to $-i\hbar\,\partial\Psi/\partial t$, we obtain

$$-\frac{\hbar^2}{2m}\Delta\Psi + V(x, y, z)\Psi = -i\hbar\frac{\partial\Psi}{\partial t} \tag{44}$$

an equation identical to the general equation (40).

We thus see that the general equation of propagation can be put in the form

$$H(x, y, z, P_x, P_y, P_z, t)\Psi = -i\hbar\frac{\partial\Psi}{\partial t} \tag{45}$$

where P_x, P_y, P_z are respectively the operators $i\hbar\,\partial/\partial x$, $i\hbar\,\partial/\partial y$ and $i\hbar\,\partial/\partial z$, which we make correspond in this way to the components of the momentum.

It is important to notice that the automatic process which we have just indicated for obtaining the wave-equation would not in general succeed if one employed curvilinear coördinates. Thus, with spherical coördinates one would not by this process obtain the correct form of the operator Δ which figures in the equation. This difficulty arises from the fact that one cannot, starting with the classical Hamiltonian function, deduce uniquely in this way, the form of the Hamiltonian operator, because a term of the form $q\mathrm{P}_q$, for example, in the classical function may give rise in the operator to such terms as $q\mathrm{P}_q$, $\mathrm{P}_q q$, $(q\mathrm{P}_q + \mathrm{P}_q q)/2$, etc., depending on the order that is adopted for the factors; and these terms are not equivalent.

To obtain the correct expression, one must first proceed to a symmetrization in p and q of the classical expression.

6. Theorem concerning group-velocity. Its reconciliation with Classical Mechanics

We now demonstrate a theorem already found in my Doctoral Thesis: the theorem of group-velocity.

Let us first recall that a wave-group is formed by a superposition of monochromatic waves having *very nearly the same* frequencies and directions of propagation. One can thus attribute to the group a frequency, a wave-length, and a direction of propagation which are approximately well determined, although the group is not rigorously equivalent to a monochromatic wave.

A wave-group may have finite dimensions, because the waves of which it is made up and which coincide in phase at the center of the wave-trains cancel each other out by interference beyond the limits of the wave-group. It is easy to see that the dimensions of such a limited wave-group must always be large in relation to the average wave-length λ_0. If, in fact, the various components coincide in phase at the center of the wave-group, and if this group is formed by a superposition of waves having wave-lengths included in the interval $\lambda_0 - \Delta\lambda \rightarrow \lambda_0 + \Delta\lambda$ with $\Delta\lambda \ll \lambda_0$, so that the components cancel each other out by interference outside of the domain occupied by the wave-group, it is then necessary that the average phase displacement of the waves of the wave-length λ_0 and $\lambda_0 \pm \Delta\lambda$ should be $\pi/2$ as one passes from the center to the boundary of the wave-group. If d is the average distance from the center to the boundary, one should thus have

$$\frac{d}{\lambda_0} - \frac{d}{\lambda_0 \pm \Delta\lambda} \simeq \frac{d\Delta\lambda}{\lambda_0^2} \simeq \frac{\pi}{2} \quad \text{or} \quad \frac{d}{\lambda_0} \simeq \frac{\pi}{2}\,\frac{\lambda_0}{\Delta\lambda} \gg 1 \qquad (46)$$

Let us now develop Rayleigh's formula, which gives the "group-velocity." In a medium with a variable index of refraction, we may write the expression for a monochromatic wave, using the approximation of Geometrical Optics in the form (23). A wave-group in this medium will then be represented by

$$\Psi = \int_{\nu_0-\Delta\nu}^{\nu_0+\Delta\nu} \mathrm{d}\nu\, a(\nu) \mathrm{e}^{2\pi i(\nu t - \varphi_1)} \qquad (\Delta\nu \ll \nu_0) \tag{47}$$

Let us put $\nu = \nu_0 + \eta$, η varying from $-\Delta\nu$ to $+\Delta\nu$. We can then write approximately,

$$\Psi = \mathrm{e}^{2\pi i[\nu_0 t - \varphi_1(x, y, z, \nu_0)]} \int_{-\Delta\nu}^{\Delta\nu} \mathrm{d}\nu\, \mathrm{e}^{2\pi i\eta\left[t - \left(\frac{\partial\varphi_1}{\partial\nu}\right)_0\right]} a(\eta) \tag{48}$$

where $(\partial\varphi_1/\partial\nu)_0$ is the derivative of φ_1, with respect to ν at $\nu = \nu_0$. In (48) the integral is a function of the parameter $t - (\partial\varphi_1/\partial\nu)_0$, and one can write

$$\Psi = \mathrm{F}\left[t - \left(\frac{\partial\varphi_1}{\partial\nu}\right)_0\right] e^{2\pi i[\nu_0 t - \varphi_1(x, y, z, \nu_0)]} \tag{49}$$

The wave-train thus behaves for a certain period of time approximately like a monochromatic wave with frequency ν_0 and having an amplitude that would be a function only of the parameter $t - (\partial\varphi_1/\partial\nu)_0$. Thus, if we move along a ray, that is, along a curve orthogonal to the surfaces $\varphi_1 = \text{const.}$, in such a way that $\mathrm{d}t - (\partial^2\varphi_1/\partial\nu\,\partial s)\mathrm{d}s$ equals zero, we will be moving with an exactly equal value of the amplitude. We can thus say that the amplitudes of the wave-group change as a whole along the rays with the velocity

$$\mathrm{U} = \frac{\mathrm{d}s}{\mathrm{d}t} = \left(\frac{\partial^2\varphi_1}{\partial\nu\partial s}\right)^{-1} \tag{50}$$

But we know that $\partial\varphi_1/\partial s = |\mathrm{grad}\ \varphi_1|$ is at every point equal to the inverse of the local wave-length $\lambda(x, y, z, \nu)$. We thus have

$$\frac{1}{\mathrm{U}} = \frac{\partial}{\partial\nu}\left(\frac{1}{\lambda}\right) = \frac{\partial\left(\frac{\nu}{\mathscr{V}}\right)}{\partial\nu} = \frac{1}{\mathscr{V}_0}\,\frac{\partial(n\nu)}{\partial\nu} \tag{51}$$

This is the famous formula of Lord Rayleigh, which gives the group-velocity U.

If the medium is homogeneous, U is independent of x, y, z. If, in addition, the medium is without dispersion $\partial n/\partial\nu = 0$, we have $\mathrm{U} = \mathscr{V}$ and the group-velocity becomes the same as the phase-velocity.

The displacement as a whole of a wave-group of velocity U is not rigorously given by the above, for in formula (48), we have neglected the terms in $\eta^2, \eta^3, \ldots$ which would contain the higher derivatives of φ_1, with respect to ν. One can see that the existence of the neglected terms has the following consequence: At the end of a sufficiently long period of time, the wave-group can no longer be considered as moving without deformation, and, eventually, it spreads out further and further into space with a diminution proportionate to its amplitude. Further on we will have to give special consideration to this progressive spreading of the wave-groups represented by a Fourier integral of type (47).

Let us now turn to the theorem of group-velocity in Wave Mechanics. For the motion of a particle moving in a static field which is derivable from the potential $V(x, y, z)$, we have found expression (36) for its wave-length, whence we derive, by recalling that $E = h\nu$,

$$\frac{\partial\left(\frac{1}{\lambda}\right)}{\partial \nu} = \frac{\partial}{\partial E}\sqrt{2m(E - V)} = \frac{m}{\sqrt{2m(E - V)}} = \frac{1}{v'} \tag{52}$$

v' being the particle's velocity corresponding to the energy E. Rayleigh's formula then gives us

$$U = v' \tag{53}$$

from which is derived the important wave-mechanical theorem of group-velocity:

If a group of Ψ waves is associated with the motion of a particle whose central frequency corresponds to the energy of the particle, the group-velocity is equal to the particle's velocity.

Let us show how this leads to a reconciliation of Wave and Classical Mechanics in the macroscopic domain. In experiments on a macroscopic scale, the fields and, consequently, the index of refraction for Ψ waves, vary little over the distance of a wave-length. In addition, since the wave-lengths are very small, we can consider wave-groups which, on our scale, almost take on the characteristics of a point. Let us consider, under these conditions, the propagation of the monochromatic wave having the central frequency ν_0 of the wave-group. Corresponding to this monochromatic wave is a family of surfaces of equal phase $\varphi_1(x, y, z, \nu_0) \doteq$ const. and the rays are the curves orthogonal to these surfaces. On the macroscopic scale, the wave-group will be analogous to a tiny bead sliding along an extremely narrow tube of rays.

Over the distance of a wave-length, it could be treated in its central portion as a monochromatic wave, and it is only along its boundaries that the interference of the various components would reduce its amplitude to zero.

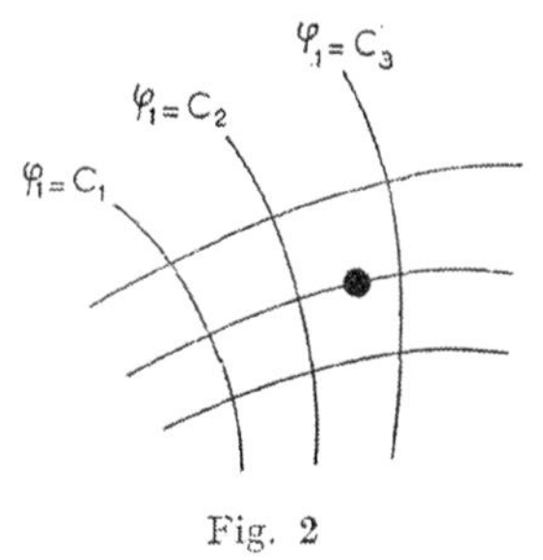

Fig. 2

The small wave-group, almost point-like on our scale, moves along a central ray with a group-velocity U which, according to (53), is equal to that of a classical point-particle describing that particular ray-trajectory. Since on the human scale we cannot distinguish the various points of the wave-group from one another, and since the particle can be apprehended only within this wave-group (because of the principle of localization, which will be studied in the next chapter), we will have the impression of being confronted by a point-particle following the motion defined by Classical Mechanics.

We thus arrive at an understanding of how Classical and Wave Mechanics may be reconciled in the case of macroscopic phenomena in which the propagation of a small group of Ψ waves can be described by the approximation of Geometrical Optics.

7. The relativistic equation of propagation in Wave Mechanics applied to a Ψ wave

We have found the equations of propagation valid in Wave Mechanics when one can neglect relativity corrections. One can seek to extend the preceding theory by taking Relativity into account.

In Relativistic Dynamics, a particle of rest mass m_0 situated outside of any possible field possesses as its energy and as its momentum

$$\mathrm{W} = \frac{m_0 c^2}{\sqrt{1-\beta^2}}, \qquad \mathbf{p} = \frac{m_0 \mathbf{v}}{\sqrt{1-\beta^2}} \qquad \beta = \frac{v}{c} \tag{54}$$

and one can separate W into two parts according to the formula

$$W = m_0 c^2 + E \qquad \text{with } E = m_0 c^2 \left[\frac{1}{\sqrt{1-\beta^2}} - 1\right] \tag{55}$$

The term $m_0 c^2$ is the rest mass energy of the particle which, following the principle of the inertia of energy, corresponds to the energy of the rest mass m_0. As for E, it is the additional energy due to the motion or "kinetic energy"—the only term taken into consideration by Classical Mechanics, which omits any consideration of the internal energy $m_0 c^2$.

When the velocity v is very small with respect to c ($\beta \ll 1$), there results immediately the classical expressions

$$E = \tfrac{1}{2} m_0 v^2, \qquad \mathbf{p} = m_0 \mathbf{v}, \qquad E = \frac{p^2}{2m_0}, \tag{56}$$

the rest mass m_0 coinciding with the constant mass m of Classical Mechanics. Classical Mechanics thus appears here as an approximation valid for $\beta \ll 1$, which is the usual case for macroscopic motions.

Whereas the classical formulas lead, in the case of the free particle, directly to the following form of the Hamiltonian function

$$E = H(x, y, z, p_x, p_y, p_z) = \frac{1}{2m}(p_x^2 + p_y^2 + p_z^2) \tag{57}$$

the relativistic formulas lead us to put

$$W = H(x, y, z, p_x, p_y, p_z) = c\sqrt{m_0^2 c^2 + p_x^2 + p_y^2 + p_z^2} \tag{58}$$

that is, to the relationship

$$\frac{W^2}{c^2} = m_0^2 c^2 + p^2 \tag{59}$$

Jacobi's equation is obtained in Classical Dynamics when one substitutes $\partial S/\partial t$ for E and $-$ grad S for $\mathbf{p}$ in the third equation (56), which gives us

$$\frac{\partial S}{\partial t} = \frac{1}{2m}\left[\left(\frac{\partial S}{\partial x}\right)^2 + \left(\frac{\partial S}{\partial y}\right)^2 + \left(\frac{\partial S}{\partial z}\right)^2\right] \tag{60}$$

Jacobi's equation in Relativistic Dynamics is obtained in the same manner, starting this time with relation (59), and has the form

$$\frac{1}{c^2}\left(\frac{\partial S}{\partial t}\right)^2 = \left(\frac{\partial S}{\partial x}\right)^2 + \left(\frac{\partial S}{\partial y}\right)^2 + \left(\frac{\partial S}{\partial z}\right)^2 + m_0^2 c^2 \tag{61}$$

To arrive at the equation of propagation in Relativistic Mechanics, it then appears natural to start with relation (59) and to make the following substitutions in it:

$$W \to -i\hbar \frac{\partial}{\partial t}, \qquad \mathbf{p} \to i\hbar \, \text{grad}$$

which leads to the equation of propagation

$$\Box \Psi + \frac{m_0^2 c^2}{\hbar^2} \Psi = 0 \tag{62}$$

Equation (62), which was put forward simultaneously by several authors at the beginning of the summer of 1926 immediately upon the heels of Schrödinger's researches, is known as the Klein-Gordon equation. In the Newtonian approximation, with expression (55) for W, one easily arrives once more, by starting with (62), at the non-relativistic form of the wave-equation [formula (40) with $V = 0$].

If we then wish to pass on to the case in which a field acts on the particle, we must give this field a relativistic form. In the case of an electron with an electrical charge e in an electromagnetic field derivable from a scalar potential V and a vector potential **A**, we will put

$$W = \frac{m_0 c^2}{\sqrt{1-\beta^2}} + eV; \qquad \mathbf{p} = \frac{m_0 \mathbf{v}}{\sqrt{1-\beta^2}} + \frac{e}{c}\mathbf{A} \tag{63}$$

and from it deduce the Hamiltonian relation:

$$\frac{1}{c^2}(W - eV)^2 = \sum_{x,y,z}\left(p_x - \frac{e}{c}A_x\right)^2 + m_0^2 c^2 \tag{64}$$

and Jacobi's equation:

$$\frac{1}{c^2}\left(\frac{\partial S}{\partial t} - eV\right)^2 = \sum_{x,y,z}\left(\frac{\partial S}{\partial x} + \frac{e}{c}A_x\right)^2 + m_0^2 c^2 \tag{65}$$

By taking into account the Lorentz relationship between the potentials $\partial V/\partial t + \text{div}\,\mathbf{A} = 0$, the usual process of passing into Wave Mechanics leads to the equation of propagation

$$\Box\Psi - \frac{2i}{\hbar}\frac{eV}{c^2}\frac{\partial \Psi}{\partial t} - \frac{2i}{\hbar}\frac{e}{c}\sum_{x,y,z} A_x \frac{\partial \Psi}{\partial x} + \frac{1}{\hbar^2}\left[m_0^2 c^2 - \frac{e^2}{c^2}(V^2 - A^2)\right]\Psi = 0 \tag{66}$$

The preceding equations, known since 1926, are valid for Ψ waves represented by an invariant wave function with a single component. The subsequent development of Wave Mechanics has shown that these equations can hold only for particles of zero spin. For particles of spin other than zero, the Ψ function must be considered as a mathematical entity having several components which undergo certain well-defined transformations under a Lorentz transformation. These components are related to each other by a system of simultaneous partial differential equations that are linear and of the first order in x, y, z, t. The number of components of the Ψ and of the corresponding equations increases as the value of the spin becomes greater. All these mathematical forms are being thoroughly studied at the present time.[3]

The simplest case is that of the electron which possesses, as probably all the other "elementary" particles do, a spin equal to $\frac{1}{2}$ in units of $\hbar$. The Wave Mechanics of the electron, discovered in 1928 by P. A. M. Dirac, is known as the "Dirac Theory". It utilizes a Ψ function with four components ψ_1, ψ_2, ψ_3, ψ_4 which obey four simultaneous partial differential equations of the first order in x, y, z, t. From these equations, one deduces that in the absence of a field, each of these ψ_k separately obeys equation (62).

We will not develop Dirac's theory here. We will return to it in the second part of the present work (Chapter XVI).

[3] On this subject one may consult the present author's work, *Théorie générale des particules à spin*, 2e éd., Gauthier-Villars, Paris, 1954.

Chapter III

FIRST PRINCIPLES RELATIVE TO THE PROBABILISTIC INTERPRETATION OF Ψ WAVES

1. The central problem in the interpretation of Wave Mechanics

From the very beginnings of the study of Wave Mechanics, the problem of the exact significance to be attributed to the Ψ wave was seen to be fraught with great difficulties. It was immediately apparent that it was not possible to consider the Ψ function as a physical quantity in the old sense—for example, as representing the vibration of some medium. The general wave equation, in fact, contains a coefficient $i = \sqrt{-1}$ so that the Ψ function is essentially complex, in contrast to the situation in the classical theory of waves and vibrations in which the use of complex functions appeared as a mathematical artifice, often convenient, but always avoidable. In addition, we will see that, for particle systems, the Ψ wave is propagated in a configuration space, which is an abstract and fictitious space.

The more the formalism of employing the Ψ wave became apparent, the more it appeared as a kind of formal and subjective representation making possible the evaluation of the probabilities of certain results of measurement. We will have occasion to show in the course of studying this probabilistic interpretation of the Ψ function that this wave-function, defined in the usual fashion as a solution of the linear equations of propagation mentioned in the preceding chapter, can by no means be considered an objective reality, but only as an element having the same subjective qualities as the probabilities it represents, an element susceptible to variations dependent upon the knowledge of the person employing it.

The overriding question, then, is to find out whether the probabilistic interpretation of the Ψ wave, which unquestionably leads to exact predictions, constitutes a "complete" representation beyond which there is no point in seeking an objective description of reality, or whether, on the contrary, the description of phenomena by the exclusive use of the Ψ wave is "incomplete" and must make room for a more profound and detailed description of physical reality. We will have occasion to return more than once to this problem.

For the moment, we will content ourselves with the exposition and study of two fundamental principles of the probabilistic interpretation of the Ψ wave, principles which came into their own as early as 1926-1927, notably as a result of the work of Max Born, the correctness of which appears beyond doubt.

2. The principle of localization or principle of interference

The first principle to become established in Wave Mechanics when the Ψ wave-function was put to use in predicting phenomena was the principle which I will call "the principle of localization", or "the principle of interference", which consists of the following:

The square of the modulus of the Ψ function measures at every point and for every instant, the probability of the particle's presence being observed at that point and instant.

The Ψ function being a complex quantity, we can always write it in the form $\Psi = ae^{\frac{i}{\hbar}\varphi}$, where a and φ are real magnitudes which we will call respectively the amplitude and phase of Ψ. If we designate by Ψ^* the complex conjugate quantity of Ψ equal to $ae^{-\frac{i}{\hbar}\varphi}$, we have

$$a^2 = \Psi^*\Psi = |\Psi|^2 \tag{1}$$

It is this real magnitude which appears in the principle of interference.

It is easy to relate the principle of interference to ideas that were well known in the older theories of light. In those theories it was always taken for granted that the intensity of the light-wave, that is, the square of its amplitude, measured at every point and at every instant the quantity of light-energy that could be captured there and then. It was this rule which permitted the exact prediction of interference and diffraction fringes. But we now know that in exchanges of energy between matter and light everything behaves as if light were composed of particles with the energy $h\nu$—the so-called "photons". If we imagine a light-wave carrying along a great number of photons, the explanation of the interference will require that the intensity of the wave shall measure at every point the density of the cloud of photons. And thus we obtain a statistical interpretation of the principle of interference applied to photons.

But this statistical interpretation must take on a probabilistic form. And indeed, light interference phenomena of the ordinary type have been achieved by using, over a very long period of time, light of an extremely low intensity—so low that there could never be more than one photon at a time in the interference apparatus. These were the basic experiments of Taylor, Dempster and Batho.[1] Now, other experiments, due to Silberstein and confirmed by those of Vavilov,[2] have shown that the image on the photographic plates is due to photoelectric effects produced locally in their sensitive layer by the successive arrival of photons. Thereafter, the only possible interpretation of Taylor-type experiments is this: Every photon reaches the interference apparatus with its wave-train, which is there subject to the interferences calculated according to wave theory; and, at the end of a very long time, when a large number of photons have reached the apparatus, it is found that those captured by the photographic plate have produced upon that plate photoelectric effects proportionate to the local intensity of the wave. One is thus led to say that the intensity of the wave must measure the probability of a photon producing an observable effect in one point in space. This probabilistic interpretation of the principle of interference makes it possible to explain the success of the experiments of Taylor, Dempster and Batho.

What has just been said concerns photons, but the extension of the principle of interference to particles of matter and to their associated waves seems inevitable, because interference and diffraction phenomena may be obtained with them also. In the case of electrons, for example—which may be easily utilized in experiments (electrons of a few hundred to a few hundred thousands of electron-volts)—the associated wave has, according to the formula $\lambda = h/p$, a wave-length of the order 10^{-8} cm to 10^{-9} cm. One should therefore be able to obtain, with electrons, phenomena analogous to those obtained with X-rays or γ-rays, which have a wave length of the same order. And that, as is well known, is what was demonstrated in 1927 by the experiments of Davisson and Germer, which were shortly thereafter continued by G. P. Thomson, Rupp, Ponte, Kikuchi, etc. These experiments showed that a cluster of free electrons having equal energy, and thus associated with a monochromatic wave, can, by being diffracted on a crystal, give rise to phenomena analogous to those observed on X-rays in experiments of the Laue-Bragg type. Moreover, Rupp was able to achieve the diffraction of electrons by means of an ordinary optical grating

[1] G. I. TAYLOR, *Proc. Cambridge Phil. Soc.*, 15 (1909) 114; DEMPSTER and BATHO, *Phys. Rev.*, 30 (1927) 644.

[2] S. I. VAVILOV, *Progrès des sciences physiques*, 16 (1936) 892.

using very oblique incidence; and, in 1940, Borsch, repeating a memorable experiment of Fresnel's on light, but this time with electrons, was able to observe the diffraction of electrons by the edge of a screen. Finally, one must recall that by means of crystals it has been possible to achieve the diffraction of a whole series of material particles other than electrons—protons, for example, various atomic nuclei, and even neutrons. All these beautiful experiments have given excellent confirmation to the general ideas of Wave Mechanics as well as a very precise quantitative verification of the formula $\lambda = h/p$. They have also given decisive support to the idea that it is correct to extend the principle of interference to all particles, since that principle is the basis of all the calculations that permit the prediction of interference and diffraction phenomena.

It is certain that the principle of interference must, for all particles, receive the probabilistic interpretation that, in the above, we were led to attribute to it in the case of photons and of light-waves. There seems no possible further doubt on the matter, since Bibermann, Suchkine and Fabrikant [3] have been able to verify that the diffraction figures obtained *with electrons* are likewise produced by the localized actions resulting from the successive arrivals of electrons on the photographic plate that records them.

3. Precise statement of the principle of interference. The probability-current

In order to define as exactly as possible the principle of interference, we note that the Ψ wave, which is a solution to a linear partial differential equation, does not possess the quality of a measurable physical magnitude, and is determined only to within a complex multiplicative factor. We can choose this factor in such a way that

$$\int d\tau \, |\Psi|^2 = 1 \qquad (2)$$

the integral being extended over all space. The choice of this arbitrary factor then permits us to "normalize" the wave function by use of relation (2), at least at a given instant. We are going to show that the Ψ function thus normalized will remain normalized at every instant t. We may thus state the principle of interference precisely by saying: *The probability that an observation will permit localization of a particle,*

[3] *Rapports à l'Académie des Sciences de l'U.R.S.S.*, 66 (1949) 185.

whose normalized *wave-function is* $\Psi(x, y, z, t)$ *within an element of volume* $d\tau$ *at the instant* t, *is equal to the expression*

$$\Psi^*(x, y, z, t)\Psi(x, y, z, t)d\tau = |\Psi(x, y, z, t)|^2 d\tau$$

In order to represent the variations of the particle's probability of presence $|\Psi|^2$ in the course of time, we will imagine a current of which the density, by hypothesis, at every point and at every instant, is given by

$$\rho(x, y, z, t) = |\Psi(x, y, z, t)|^2 \tag{3}$$

We define the motion of this current by assuming that the velocity at point x, y, z at the instant t has the form

$$\mathbf{v} = \frac{1}{|\Psi|^2}\frac{\hbar}{2im}[\Psi \operatorname{grad} \Psi^* - \Psi^* \operatorname{grad} \Psi] = -\frac{1}{m}\operatorname{grad} \varphi \tag{4}$$

Now, the functions Ψ and Ψ^* obey respectively equation (41) of Chapter II (p. 20) and the conjugate equation, whence we easily obtain

$$\Psi^* \Delta\Psi - \Psi\Delta\Psi^* = \frac{2im}{\hbar}\frac{\partial}{\partial t}(\Psi^*\Psi) = \frac{2im}{\hbar}\frac{\partial \rho}{\partial t} \tag{5}$$

We can thus note

$$\frac{\partial \rho}{\partial t} = \frac{\hbar}{2im}[\Psi^*\Delta\Psi - \Psi\Delta\Psi^*] = -\frac{\hbar}{2im}\sum_{x,y,z}\frac{\partial}{\partial x}\left(\Psi\frac{\partial \Psi^*}{\partial x} - \Psi^*\frac{\partial \Psi}{\partial x}\right) \tag{6}$$

or according to (4)

$$\frac{\partial \rho}{\partial t} + \operatorname{div}(\rho\mathbf{v}) = 0 \tag{7}$$

This equation, well known in Hydrodynamics as the continuity equation, shows that the current of density ρ is conserved despite the passage of time, that is to say, the integral $\int d\tau\, |\Psi|^2$ remains constant. The normalization of the Ψ thus definitely has a permanent character.

4. The Heisenberg uncertainty relations

Classical Mechanics asserted that it was possible to assign a well-determined position and velocity to a particle. In other words, at every instant clearly determined values were attributed to the particle's coordinates x, y, z and to its energy E as well as to the momentum $\mathbf{p}$. We will demonstrate that in Wave Mechanics, at least when we consider exclusively the continuous wave-function Ψ, this assertion can no longer be maintained.

Let us study the simple case of uniform rectilinear motion in a completely field-free region. We know that to the motion with energy E and momentum **p**, where the direction cosines are α, β, γ, there corresponds the propagation of a plane monochromatic wave

$$\Psi = ae^{\frac{i}{\hbar}[Et - \sqrt{2mE}(\alpha x + \beta y + \gamma z)]} \tag{8}$$

with frequency E/h and wave-length h/mv, at least in the Newtonian approximation. This plane monochromatic wave thus corresponds to a well-determined state of motion; but it gives no indication as to the position of the particle, because it has the same amplitude at every point in space. The probability of the particle's presence, $|\Psi|^2$, is thus the same for all points.

But, instead of being a plane monochromatic wave, the solution Ψ of the wave-equation that corresponds to the state of the particle may be a superposition of plane monochromatic waves representing a wave-train of limited dimensions. Then the quantity $|\Psi|^2$ will differ from zero only in a limited region of space, and according to the principle of interference, the presence of the particle can be detected only in that region. The uncertainty about the position is thus smaller than in the case of the plane monochromatic wave. On the other hand, if to each monochromatic component of frequency ν and wave-length λ that is present in the spectral make-up of the Ψ, we associate the state of motion defined by

$$E = h\nu, \quad p_x = \alpha\frac{h}{\lambda}, \quad p_y = \beta\frac{h}{\lambda}, \quad p_z = \gamma\frac{h}{\lambda} \tag{9}$$

we can no longer attribute a well-defined state of motion to the particle. Passing from the case of the plane monochromatic wave to that of the limited wave-train, we have thus seen the uncertainty diminish in respect to position; but, correspondingly, we have seen the uncertainty increase in respect to the state of motion.

We may now take up the limiting case of a wave-train of infinitesimally small dimensions. It then becomes necessary to introduce, in order to represent this wave-train analytically, a superposition of monochromatic waves having all possible frequencies, wave-lengths and directions. This limiting case, like that of the plane monochromatic wave, allows a well-determined localization of the particle, but leaves us completely in ignorance concerning its state of motion.

To sum up: The better the position of the particle is defined, the greater is the uncertainty about its state of motion, and vice-versa. The qualitative observation may be more precisely defined by an exact analysis using the Fourier integral of the representation of a train of waves. In this way it can be shown that, if our knowledge of the Ψ wave allows the uncertainties Δx, Δy, Δz with respect to the particle's coordinates, and the uncertainties Δp_x, Δp_y, Δp_z with respect to the components of its momentum, we have between these uncertainties the relations

$$\Delta x \Delta p_x \geq h, \qquad \Delta y \Delta p_y \geq h, \qquad \Delta z \Delta p_z \geq h \tag{10}$$

valid in order of magnitude. These are the "Heisenberg uncertainty relations". They show that the product of the uncertainty with respect to one coordinate and the uncertainty of the corresponding component of the momentum is always of order h.

To the relations of (10) we add a fourth uncertainty relation resulting from them

$$\Delta t \, \Delta \mathrm{E} \geqq h \tag{11}$$

in order of magnitude. The interpretation of relation (11) is this: If a measurement, which permits assigning a definite value to the energy E of a particle, lasts for a time Δt, the value obtained for E is uncertain by an amount ΔE which is greater than or equal to $h/\Delta t$.[4]

5. The principle of spectral decomposition (Born)

In the reasoning just presented we have implicitly admitted a principle that we must now state explicitly. This principle, which was deduced from the earliest applications of Wave Mechanics, and which Born was the first to apprehend clearly, is as follows:

If the Ψ wave is formed by the superposition of a certain number of plane monochromatic waves, then to each of its components there corresponds a possible state of motion of the particle, that is to say, an observation may justify the attributing of this state of motion to the particle.

In exact terms, one can say with Born:

If the Ψ wave is formed by a superposition of plane monochromatic waves forming a discontinuous series, that is, if

$$\Psi = \sum_k c(\mathbf{p}_k) e^{\frac{i}{\hbar}[\mathrm{E}_k t - \mathbf{p}_k \cdot \mathbf{r}]} \tag{12}$$

[4] For further remarks on the Heisenberg uncertainty principle in the light of the theory of the double solution see ref. [17] pp. 93–95.

where E_k *is the energy corresponding to the momentum* $\mathbf{p}_k$, *the probability that an observation will lead one to assign the state with the momentum* $\mathbf{p}_k$ *to the particle is given by* $|c(\mathbf{p}_k)|^2$. *If, on the other hand, the* Ψ *wave is formed by a superposition of plane monochromatic waves forming a continuous spectrum (which is the case with usual wave-trains), that is, if*

$$\Psi = \int d\mathbf{p}\, c(\mathbf{p}) e^{\frac{i}{\hbar}[Et - \mathbf{p}\cdot\mathbf{r}]} \qquad (d\mathbf{p} = dp_x dp_y dp_z) \tag{13}$$

the probability that an observation will lead one to assign to the particle a momentum included between $\mathbf{p}$ *and* $\mathbf{p} + d\mathbf{p}$ *is given by* $|c(\mathbf{p})|^2 d\mathbf{p}$, *remembering that* $c(\mathbf{p})$ *is given as a function of* $\Psi(\mathbf{r}, t)$ *by the Fourier inversion formula:*

$$c(\mathbf{p}) = \frac{1}{h^3} \int d\mathbf{r}\, \Psi(\mathbf{r}, t) e^{-\frac{i}{\hbar}[Et - \mathbf{p}\cdot\mathbf{r}]} \qquad (d\mathbf{r} = dx dy dz) \tag{14}$$

It may thus be said that the probability of each state is measured by the "intensity" of the corresponding spectral component. The states of motion that are not present in the Fourier expansion of the Ψ wave function have a zero probability. This conclusion may be considered as forming the basis of the theory of quantized states in Wave Mechanics.

6. Comments on the foregoing conclusions

The conclusions just presented permit us to state precisely the significance of the Ψ wave. It does not constitute a physical quantity in the classical sense; it is only an instrument for prediction—more precisely, an instrument for predicting probability. Its form arises from previous observations, the results of which have become known to the observer and have given him data on the state of the particle. With these initial data as the point of departure, and as long as no new information comes to the knowledge of the observer, the Ψ function's form evolves in conformity with the wave equation.

Although this evolution of the Ψ wave is entirely determined, there does not result from it—as we will see—a rigorous prediction of future observations. Indeed, the knowledge of the Ψ wave does not permit us to say what value of a given quantity will be observed at the moment of a new observation; it permits us to say only what the values are that may be found for the quantity and their respective probabilities.

Each time a new observation provides us with new information about the particle, the form of the Ψ wave is thereby modified. This becomes understandable only if the Ψ wave is merely a representation of our present knowledge concerning the particle and not the representation of an objective reality. As Schrödinger said recently, there is something "psychological" about the Ψ wave.

We will see that observations made simultaneously during the course of one and the same experiment can never supply us with more precise knowledge concerning the magnitudes which characterize the particle than is permitted by the Heisenberg uncertainty relations. If we measure with precision the values of certain magnitudes, the values of the canonically conjugate magnitudes remain totally unknown to us. There are thus "maximal" experiments which supply us with the greatest possible knowledge about the particle compatible with the Heisenberg relations. If the possibility existed of finding experiments that would permit us to know precisely all the magnitudes attached to the particle, the Heisenberg uncertainty relations would no longer be satisfied; and so we conclude from the reasoning previously presented that, after an experiment of this sort, we could no longer represent the state of our knowledge by a Ψ wave; but subtle analyses, due chiefly to Bohr and Heisenberg, have shown that no experiment of this sort can ever be performed, by very reason of the existence of the quantum of action h.

These are the conclusions that one reaches by adhering exclusively to the point of view that is considered orthodox at the present time and according to which the Ψ wave furnishes us with a complete description of the state of a particle. All these conclusions appear, moreover, thoroughly verified, and we shall have to ask ourselves how their success might be interpreted if we adopted a point of view different from the orthodox one.

Incidentally, we may also point out that the distribution of probability supplied by the principle of interference and that of spectral decomposition for two canonically conjugate magnitudes x and p_x leads *rigorously* to the relations between the "dispersions"

$$\sigma_x = \sqrt{\overline{(x - \bar{x})^2}}, \qquad \sigma = \sqrt{\overline{(p_x - \bar{p}_x)^2}} \tag{15}$$

the relation

$$\sigma_x \sigma_{p_x} \geqq \frac{\hbar}{2}, \ldots \tag{16}$$

This constitutes a precise *form* of the *uncertainty relations*, but contrary to what is sometimes said, it is not exactly equivalent to the qualitative form

$$\Delta x \Delta p_x \geq h \qquad \text{in order of magnitude.} \tag{17}$$

Indeed, cases may arise in which the relation (16) provides no exact information (if one of the dispersions is infinite), whereas the qualitative relation (17) always possesses a practical value for the experimenter.

7. The theory of probability-current in relativistic terms

It is easy to transpose the theory of the probability-current into the terms of the relativistic wave equation

$$\Box \Psi + \frac{m_0^2 c^2}{\hbar^2} \Psi = 0 \tag{18}$$

One has only to put $\Psi = a e^{\frac{i}{\hbar}\varphi}$ and to define the density and the velocity of the probability-current by

$$\rho = a^2 \frac{\partial \varphi}{\partial t} \qquad \mathbf{v} = -c^2 \operatorname{grad} \varphi \bigg/ \frac{\partial \varphi}{\partial t} \tag{19}$$

in order to verify the equation of continuity

$$\frac{\partial \rho}{\partial t} + \operatorname{div} \rho \mathbf{v} = 0. \tag{20}$$

This is easily verified by substituting the expression for Ψ into (18) and by cancelling out the pure imaginary terms in the resultant equation.

If, instead of a free particle, we considered an electron subjected to an electromagnetic field derivable from the potentials V and **A**, equation (18) would have to be replaced by equation (66) of Chapter II (p. 27). It would then be seen that, in order to end up with the continuity equation (20), one must put, instead of (19), the formula

$$\rho = a^2 \left(\frac{\partial \varphi}{\partial t} - eV \right); \qquad \mathbf{v} = -c^2 \frac{\operatorname{grad} \varphi + \frac{e}{c} \mathbf{A}}{\frac{\partial \varphi}{\partial t} - eV} \tag{21}$$

We will come upon the expression for **v** again under the name of a "guidance formula" in the attempted development of a causal theory, which forms the second part of the present work.

One can easily verify that (21) may be written in the form

$$\rho = -\frac{\hbar}{2i}\left(\Psi\frac{\partial\Psi^*}{\partial t} - \Psi^*\frac{\partial\Psi}{\partial t}\right) - |\Psi|^2 eV$$

$$\rho\mathbf{v} = \frac{\hbar c^2}{2i}(\Psi\,\mathbf{grad}\,\Psi^* - \Psi^*\,\mathrm{grad}\,\Psi) - |\Psi|^2 ec\mathbf{A} \qquad (22)$$

These formulas make it possible to verify that ρc and the three components of $\rho\mathbf{v}$ form the four components of a space-time quadrivector.

It is likewise easy to verify that in the Newtonian approximation (where $(\partial\varphi/\partial t) \simeq m_0 c^2$), formulas (19) bring us back to the formulas of paragraph 3, ρ being defined only within a multiplicative constant.

The density-current quadrivector $(\rho, \rho\mathbf{v})$ of Dirac's theory will be studied later. (Part II, Chapter 16).

Chapter IV

THE WAVE MECHANICS OF SYSTEMS OF PARTICLES

1. The Classical Dynamics of point-mass systems

Up to this point we have considered only the case of a single particle situated in a known field of force. How can we generalize the results set forth in the preceding pages to apply to an ensemble of particles acting upon each other? To see how this can be done it will first be necessary to recall the main lines of the Classical Dynamics of point-mass systems.

Let us consider a system made up of N particles. The mass of the ith particle is m_i, its coordinates are x_i, y_i, z_i. The kinetic energy of the system is

$$T = \tfrac{1}{2}\sum_i m_i(\dot{x}_i^2 + \dot{y}_i^2 + \dot{z}_i^2) \quad \text{with } \dot{x}_i = \frac{dx_i}{dt}, \ldots \tag{1}$$

The conjugate Lagrange momenta are

$$p_{x_i} = m_i \frac{dx_i}{dt}, \quad p_{y_i} = m_i \frac{dy_i}{dt}, \quad p_{z_i} = m_i \frac{dz_i}{dt}. \tag{2}$$

The potential energy of the system $V(x_1 \ldots, z_N, t)$ is composed of two kinds of terms:

1. those which express the mutual action of the particles and which are assumed to depend only on the distances between the particles; they are of the form

$$V_{ij} = V_{ij}[\sqrt{(x_i - x_j)^2 + (y_i - y_j)^2 + (z_i - z_j)^2}]; \tag{3}$$

2. those that express the possible action of an external field on each of the particles; they are of the form $V_i(x_i, y_i, z_i, t)$.

The Hamiltonian giving the expression for the energy as a function of the coordinates and momenta is

$$H(x_1, \ldots, z_N, t) = \sum_{i=1}^{N} \frac{1}{2m_i}(p_{x_i}^2 + p_{y_i}^2 + p_{z_i}^2) + V(x_1, \ldots, z_N, t). \tag{4}$$

If the external field is not dependent upon time or is zero, V does not contain t and we know that then H has a constant value E in the course of motion (conservative system).

Jacobi's theory may be extended to such systems. Jacobi's equation for the system is

$$\sum_{k=1}^{N} \frac{1}{2m_k}\left[\left(\frac{\partial S}{\partial x_k}\right)^2 + \left(\frac{\partial S}{\partial y_k}\right)^2 + \left(\frac{\partial S}{\partial z_k}\right)^2\right] + V(x_1, \ldots, z_N, t) = \frac{\partial S}{\partial t}. \quad (5)$$

If one succeeds in finding a complete integral for this equation containing 3N non-additive arbitrary constants $\alpha_1, \alpha_2, \ldots, \alpha_{3N}$, one of the possible motions of the system may be obtained by writing

$$\frac{\partial S}{\partial \alpha_i} = a_i \quad (i = 1, 2, \ldots, 3N) \quad \text{with } S = S(x_1 \ldots z_N, t, \alpha_1 \ldots \alpha_{3N}) \quad (6)$$

where the a_i are new arbitrary constants, and the Lagrange momenta will be given by the formulas

$$p_{x_i} = -\frac{\partial S}{\partial x_i}, \qquad p_{y_i} = -\frac{\partial S}{\partial y_i}, \qquad p_{z_i} = -\frac{\partial S}{\partial z_i} \quad (7)$$

In the important special case where external actions are independent of time (or zero), V does not contain t and one may envisage solutions of (5) of the form

$$S = Et - S_1(x_1 \ldots z_N, \alpha_1 \ldots \alpha_{3N-1}, E) \quad (8)$$

with 3N constants $\alpha_1, \alpha_2, \ldots, \alpha_{3N-1}$, E. One is then brought to consider the "separated" Jacobi equation

$$\sum_{k=1}^{N} \frac{1}{2m_k}\left[\left(\frac{\partial S_1}{\partial x_k}\right)^2 + \left(\frac{\partial S_1}{\partial y_k}\right)^2 + \left(\frac{\partial S_1}{\partial z_k}\right)^2\right] + V(x_1, \ldots, z_N) = E \quad (9)$$

and to seek a complete integral for it containing 3N arbitrary constants $\alpha_1, \ldots, \alpha_{3N-1}$, E. The equations of the motion are first

$$\frac{\partial S_1}{\partial \alpha_i} = a_i \qquad (i = 1, \ldots, 3N - 1) \quad (10)$$

which give the trajectory of the representation point of the system in the configuration space whose dimensions are the variables $x_1, \ldots, z_N$. Then one has the equation for time

$$\frac{\partial S_1}{\partial E} = t - t_0. \quad (11)$$

Finally

$$p_{x_i} = \frac{\partial S_1}{\partial x_i}, \qquad p_{y_i} = \frac{\partial S_1}{\partial y_i}, \qquad p = \frac{\partial S_1}{\partial z_i}. \tag{12}$$

As in the case of a single point mass, Jacobi's equation permits us to define "classes" of motion of the representation point of the system in configuration space, each class corresponding to a function $S_1(x_1, \ldots, z_N, \alpha_1, \ldots, \alpha_{3N-1}, E)$ with values given for the primary constants $\alpha_1, \ldots, \alpha_{3N-1}, E$, and the various motions of a single class being characterized by the value of the secondary constants $a_1 \ldots, a_{3N-1}, t_0$.

2. The Wave Mechanics of systems of particles

To obtain a Wave Mechanics for particle systems, Schrödinger in his work of 1926 introduced the idea that it was necessary to consider the propagation of Ψ waves in the configuration space of the system, such that the Geometrical Optics of this wave propagation will bring us back to the theory of Jacobi set forth in the preceding section.

So one allows the equation of the propagation of Ψ waves in configuration space to be obtained by the same automatic procedure which succeeded in the case of a single particle. Thus we start from the classical Hamiltonian expression $H(x_1, \ldots, z_N, p_{x_1}, \ldots, p_{z_N}, t)$ for the appropriate energy of the system under consideration, and we transform this function into an operator by replacing $p_{x_k}, p_{y_k}, p_{z_k}$ with the operators

$$P_{x_k} = i\hbar \frac{\partial}{\partial x_k}, \qquad P_{y_k} = i\hbar \frac{\partial}{\partial y_k}, \qquad P_{z_k} = i\hbar \frac{\partial}{\partial z_k}. \tag{13}$$

We thus obtain the Hamiltonian operator $H(x_1, \ldots, z_N, P_{x_1}, \ldots, P_{z_N}, t)$, and we adopt as the equation for the propagation of Ψ waves in configuration space

$$-i\hbar \frac{\partial \Psi}{\partial t} = H(x_1, \ldots z_N, P_{x_1} \ldots P_{z_N}, t)\Psi. \tag{14}$$

With $x_1, \ldots, z_N$ being the rectilinear coördinates of the N particles, we thus find

$$\sum_{k=1}^{N} \frac{1}{m_k}\left(\frac{\partial^2\Psi}{\partial x_k^2} + \frac{\partial^2\Psi}{\partial y_k^2} + \frac{\partial^2\Psi}{\partial z_k^2}\right) - \frac{2}{\hbar^2} V(x_1 \ldots z_N, t)\Psi = \frac{2i}{\hbar}\frac{\partial \Psi}{\partial t}. \tag{15}$$

For $N = 1$, we quite naturally obtain once more the equation valid for the case of a single particle.

For the conservative systems ($\partial V/\partial t = 0$), one can consider monochromatic solutions that depend on time only through the factor $e^{\frac{i}{\hbar}Et}$, and equation (15) will be written

$$\sum_{k=1}^{N} \frac{1}{m_k} \Delta_k \Psi + \frac{2}{\hbar^2} [E - V(x_1, \ldots, z_N)] \Psi = 0. \tag{16}$$

If, in a region of configuration space V varies only slightly over the distance of a local wave-length, Geometrical Optics is valid and the wave has the approximate form

$$\Psi = a e^{\frac{i}{\hbar}[Et - S_1]} \tag{17}$$

a being a slowly varying amplitude whose derivatives are very small with respect to the variations of S_1. By substituting (17) in (16), we see that S_1 must be a solution of Jacobi's equation (9)—a fact that establishes the link with Classical Mechanics.

An interesting special case is that in which the particles of the system do not interact with each other. One can then consider them either in isolation or as making up a system. The V function is reduced to the terms $V_i(x_i, y_i, z_i, t)$ which express the action of the external field on the various particles, and equation (15) is written

$$\sum_{k=1}^{N} \frac{1}{m_k} \Delta_k \Psi - \frac{2}{\hbar^2} \sum_{k=1}^{N} V_k(x_k, y_k, z_k, t) \Psi = \frac{2i}{\hbar} \frac{\partial \Psi}{\partial t}. \tag{18}$$

If we put

$$\Psi(x_1 \ldots z_N, t) = \Psi_1(x_1, y_1, z_1, t) \ldots \Psi_N(x_N, y_N, z_N, t) \tag{19}$$

we find that the equation of the system breaks up into N equations of the type

$$\frac{1}{m_k} \Delta_k \Psi_k - \frac{2}{\hbar^2} V_k(x_k, y_k, z_k, t) \Psi_k = \frac{2i}{\hbar} \frac{\partial \Psi_k}{\partial t} \tag{20}$$

and we see that each particle may be considered separately.

Nevertheless, form (19) of the solutions is not the most general one; the equation of propagation likewise permits solutions that are made up of any linear combination whatsoever of the products $\prod_k \Psi_k(x_k, y_k, z_k, t)$. These combinations are appropriate to the description of cases where the particles no longer have individual states as a result of their previous interactions. There will be occasion to come come back later to these "correlated" states resulting from previous interactions. Solutions of type (19) are applicable only when the states of the particles are independent.

3. The probabilistic interpretation of the Wave Mechanics of systems of particles

It is easy to transpose the principle of interference to the case of systems of particles. We then state it by saying: *If the state of a system of particles is represented in configuration space by the wave-function* $\Psi(x_1, \ldots, z_N, t)$, *the probability that an observation will permit the localization of are presentation point of the system, at the time t, in the element of volume of configuration space*

$$d\tau = dx_1\, dy_1\, dz_1 \ldots dx_N\, dy_N\, dz_N$$

is equal to $|\Psi(x_1, \ldots, z_N, t)|^2 d\tau$.

If there is only one particle, we are obviously back at the previously studied form of the principle of interference. For non-interacting particles that have *never* interacted (independent particles), one may adopt expression (19) for Ψ and consequently write

$$|\Psi|^2 d\tau = |\Psi_1(x_1, y_1, z_1, t)|^2 dx_1\, dy_1\, dz_1 \ldots |\Psi_N(x_N, y_N, z_N, t)|^2 dx_N\, dy_N\, dz_N. \tag{21}$$

The probability that the representation point of the system will be found in the element of volume $dx_1 \ldots dz_N$ of configuration space is then the product of the individual probabilities that the first particle will be found in an element of volume dx_1, dy_1, dz_1, . . ., the Nth in the element of volume $dx_N\, dy_N\, dz_N$. This result is in agreement with the theory of composite probabilities, for the presences of the various particles in the different elements of physical space are here independent events; and we can plainly see why, in this case, the ordering of the Ψ function must have the form (19).

For the magnitude $|\Psi|^2$ to give, in absolute value, the probability of the presence of the representation point of the system in the element of volume $d\tau$ of configuration space, we must normalize the Ψ by putting

$$\underbrace{\int \cdot \cdot \int}_{3N} d\tau |\Psi|^2 = 1 \tag{22}$$

which determines Ψ to within a constant factor $e^{i\alpha}$.

We still have to demonstrate that the normalization (22) valid at an instant t is maintained in time. In order to do this, let us consider an imaginary current defined in configuration space by the formulas

$$\rho = |\Psi|^2$$
$$\rho \mathbf{v}_k = \frac{\hbar}{2im_k} [\Psi \operatorname{grad}_k \Psi^* - \Psi^* \operatorname{grad}_k \Psi] = -\frac{1}{m_k} |\Psi|^2 \operatorname{grad}_k \varphi \tag{23}$$

with $\Psi = ae^{\frac{i}{\hbar}\varphi}$, $\mathbf{v}_k$ having as components dx_k/dt, dy_k/dt, and dz_k/dt and grad_k having as components $\partial/\partial x_k$, $\partial/\partial y_k$, and $\partial/\partial z_k$.

By multiplying the equation of propagation (15) by Ψ^*, the conjugate equation by Ψ, and by subtracting, one then obtains

$$\sum_{k=1}^{N} \frac{1}{m_k} [\Psi^* \Delta_k \Psi - \Psi \Delta_k \Psi^*] = \frac{2i}{\hbar} \frac{\partial}{\partial t} (\Psi^* \Psi) \tag{24}$$

from which one can deduce with the aid of the definitions (23)

$$\frac{\partial \rho}{\partial t} + \sum_{k=1}^{N} \operatorname{div}_k (\rho \mathbf{v}_k) = 0 \tag{25}$$

an equation that is the generalization of the continuity equation of hydrodynamics for an imaginary current in motion in a space of 3N dimensions. The current of probability is thus preserved in its motion in configuration space, and the normalization of Ψ has a permanent character.

The principle of spectral decomposition is here stated as it applies to a single particle. If the system is conservative, the Ψ wave may always be represented by a superposition of monochromatic waves, and the intensity of each spectral component gives the probability that an observation will permit the assigning to the system the corresponding energy of the component.

By studying the representation of a train of Ψ waves in configuration space by means of a Fourier integral, and by comparing the distributions of probability for x_k and p_{x_k}, we would again come out with the qualitative relations of uncertainty

$$\Delta x_k \Delta p_{x_k} \geqq h \quad \text{in order of magnitude} \tag{26}$$

and the exact relation between the dispersions

$$\sigma_{x_k} \sigma_{p_{x_k}} \geqq \frac{\hbar}{2}. \tag{27}$$

In the preceding theory, we assume that the particles are free to move about in all space (a system without constraints), and we have used the rectangular Cartesian coordinates of these particles in locating the system. If we wish to use curvilinear coordinates—which is normal in the case of constrained systems where the number of degrees of freedom is less than 3N, the foregoing calculations must be worked out somewhat differently. But we will not pursue that development here.

4. Systems of particles having the same physical nature

The case of a system formed by particles of the same physical nature presents very important peculiarities which we will now summarize. This is especially the case of atoms when one can consider the nucleus as a simple center of forces surrounded by electrons.

What characterizes a system containing particles of the same physical nature is the fact that its state should be in no wise modified if the role of any two of these identical constituents is permuted. Two particles of the same sort, two electrons, for example, are so much alike that it is impossible to assign any individuality to them. One of the outstanding results of Quantum Physics was the revelation of this "indistinguishability" of particles of the same nature.

It must thus be admitted that every observable magnitude, such as $|\Psi|^2$, must be unaffected by any permutation whatsoever of the particles. And this leads us to restrict the possible form of wave-functions. Since the interactions of particles are always symmetrical functions of their coordinates, if one has found a solution

$$\Psi_1(x_1, y_1, z_1, \ldots, x_i, y_i, z_i, \ldots, x_k, y_k, z_k, \ldots, x_N, y_N, z_N, t) = \Psi_1$$

of the wave equation, the function obtained by permuting the particles i and k, that is

$$\Psi_2(x_1, y_1, z_1, \ldots, x_k, y_k, z_k, \ldots, x_i, y_i, z_i \ldots, x_N, y_N, z_N, t) = \Psi_2$$

is still a solution, as is any linear combination of the two solutions thus obtained in the form

$$a\Psi_1 + b\Psi_2 = \Psi.$$

If then we write that the $|\Psi|^2$ corresponding to this linear combination is unaffected by the permutation of the particles i and k, we arrive at the condition $b = \pm a$, and one can likewise verify that every observable magnitude constructed with the help of this Ψ is equally unaffected by this permutation if $b = \pm a$.

In order to state these results in a general form, let us introduce the following definitions: A wave-function will be called "symmetrical" if it is unchanged under the permutation of two particles of the same nature. A wave-function will be called "antisymmetrical" if, when such a permutation is performed, it changes sign without changing its absolute value. We can then state the result achieved in the following form: *The Ψ wave-function of a system which contains particles of the same nature must be either symmetrical or antisymmetrical in relation to the ensemble of the particles composing the system.*

It can be shown that there always exists an infinite number of symmetrical and an infinite number of antisymmetrical states. One finds that for each sort of particle only one or the other of these two categories is realizable in nature. In order to justify this hypothesis one demonstrates—by relying always on the symmetry of the interactions—that, if the system is initially in a symmetrical state, it can undergo transitions only to another symmetrical state; and, if it is initially in an antisymmetrical state, it can undergo transitions only to another antisymmetrical state. Thus the symmetrical systems, on one hand, and the antisymmetrical systems on the other, form closed ensembles and cannot be combined with each other.

Thus one would expect that, depending on the physical nature of the similar particles, only one of the two categories of states is to be found in nature. Experiment has confirmed this hypothesis by showing that photons, α particles, and certain atomic nuclei are particles with symmetrical states, whereas electrons, protons, neutrons, and certain other atomic nuclei are particles with antisymmetrical states.

If one studies the statistics that are valid for ensembles of particles of one and the other category, one sees that these statistics are different from each other, and that both groups of statistics differ from those of classical Boltzmann-Gibbs statistics. The particles with symmetrical states obey statistics called "Bose-Einstein statistics", and thus they are frequently referred to as "bosons". The particles with antisymmetrical states obey statistics called "Fermi-Dirac statistics", and are correspondingly referred to as "fermions". Fermions obey Pauli's principle, according to which, in an assemblage of particles, no two constituents can be found in the same individual state. Indeed, if we had two constituents playing exactly the same role in an antisymmetrical state of the system, the permutation of these two constituents —which must change the sign of the Ψ function—ought also to produce no change in the system. The wave-function, since it must remain the same and simultaneously change its sign, would necessarily be zero—which means that the state under consideration cannot really exist.

Let us note, moreover, that a more complete study shows that, to apply Pauli's principle correctly, we must take spin into account—which will not be done here.

5. Remarks on the Wave Mechanics of systems of particles

The Wave Mechanics of systems of particles as we have just set forth, following Schrödinger, is an essentially non-relativistic theory because it assumes that the interactions can be represented at every instant by functions of the actual separation distances of the particles, whereas in a relativistic theory of interactions, these interactions are propagated at a finite velocity, which introduces retardation of one sort or another. A relativistic Wave Mechanics of the systems cannot be developed along the lines we have indicated, and only recently has there been any attempt to construct such a Mechanics within the framework of Quantum Field Theory (works by Tomonaga, Schwinger, Feynman, etc.). Let us simply emphasize the fact that the theory set forth above is valid only for the Newtonian approximation.

Schrödinger's idea of identifying the Ψ wave of a system in configuration space at first shocked me very greatly, because, configuration space being a pure fiction, this conception deprives the Ψ wave of all physical reality. For me the wave of Wave Mechanics should have evolved in three-dimensional physical space. The numerous and brilliant successes that resulted from adopting Schrödinger's point of view obliged me to recognize its value; but for a long time I remained convinced that the propagation of the Ψ wave in configuration space was a purely imaginary way of representing wave phenomena which, in point of fact, take place in physical space. We will see in the second part of the present work (Chapter XII) how, from 1927 on, I had sought to develop this approach within the framework of the theory of the Double Solution.

Chapter V

A GENERAL VIEW OF THE PROBABILISTIC INTERPRETATION OF WAVE MECHANICS

1. General considerations

The development of Wave Mechanics has focussed attention on the influence which operations of measurement have on our knowledge of physical reality at the atomic level. In this field one may call "measurement" every operation made with the aid of an appropriate macroscopic device that allows us to attribute a certain value to one of the quantities characterizing a particle or a system on the atomic level.

Classical Physics implicitly assumed that one could, with the help of a sufficiently refined experimental technique, indefinitely diminish the perturbation that a measuring operation may cause in the state of things existing prior to the measurement, so that, barring experimental errors, every measurement would exactly represent the situation existing both before and after the measurement was made. Quantum Physics has noted that the existence of the quantum of action did not allow the indefinite diminution of the perturbation that a measurement produces in the prior situation. The minimum residual perturbation, which is insignificant on a large scale, can no longer be neglected on a small scale. This is what is shown by measurements of the sort known as the "Heisenberg microscope". Thus, the measurement of a quantity does not generally reveal a situation that existed prior to the measurement, but a situation brought about, in a way, by the very act of measurement. In general, one cannot say that the measured value of the quantity already existed prior to the measurement.

Moreover, the analyses of Bohr and Heisenberg, about which we have already spoken, prove that no measuring operation can with precision simultaneously furnish the values of two canonically conjugate quantities. Every attempt to measure them simultaneously ultimately always leaves intact the uncertainties Δq and Δp of their values, such that the relation $\Delta q \Delta p \geqq h$ shall be satisfied. Thus, an operation of measurement, even when carried out with all the precision that the existence of the quantum of action (maximal measurement) permits, still leaves unknown the value of at least half of the physical quantities which characterize the particle or the system being studied.

Even were we to return to a causal interpretation of Wave Mechanics, it seems that all of the ideas we have just enumerated would still have to remain intact. But the probabilistic interpretation now generally accepted attributes a very special significance to these ideas. According to that interpretation, in the state of things existing prior to the measurement, a physical quantity would not generally have a well determined value, but only possible values—that is, values that the measuring operation *can* extract from the prior state of things. Yet the knowledge of the Ψ function before measurement should allow us to say what the possible values of a quantity, as well as the respective probabilities of these possible quantities, are. The Ψ function being known, the quantities have only "potential" values, and all that one can state definitely is the "distribution of probability" for these potential values.

It is possible to illustrate these affirmations by recourse to the principle of interference and the principle of spectral decomposition. For a limited train of Ψ waves, the coordinates of the particle have as possible values those which correspond to all the points within a wave-train; the particle is said to be present "in the potential state" at every point in the wave-train. The distribution of probability of the various possible positions within the wave-train is given by the corresponding value of $|\Psi|^2$ at the instant under consideration. Similarly, if the Ψ wave is formed by a superposition of plane monochromatic waves of the form $\Psi = \sum_k c_k \Psi_k$, the particle, prior to measurement, is in a "potential state" in all the states of motion corresponding to the k indices—the distribution of probability being given by $|c_k|^2$. If we measure either the position or the momentum, we obtain one of the possible values—the *a priori* probability of the result being given by the corresponding distribution of probability.

Such is the point of view of the prevailing probabilistic interpretation. There is another point of view that we will have occasion to refer to again. It consists in allowing the quantities prior to measurement to possess precisely determined values, which are in general unknown. Measurement would disturb the situation by substituting for the initial values of the measured quantities one or the other of the values anticipated by the usual theory, and with corresponding probability.

2. An analysis of the role of operations of measurement in Quantum Physics

Whatever the interpretation we may adopt, it is of interest to form some picture of the role of measurement in Quantum Physics. In order to do this, we will consider the case of photons and light-waves, to be followed immediately by transposition to the case of material particles and their associated waves.

Let us consider any light-wave. We can "decompose" this wave by viewing it from various standpoints. If we consider it from the point of view of frequency and direction of propagation, we can "decompose" the wave spectrally in a finite or infinite series of plane monochromatic waves. This is accomplished analytically by developing the wave-function in a series or by means of a Fourier integral. If the wave is directed towards an apparatus that separates the monochromatic components—a grating, for example—one obtains each Fourier component separately, along with its appropriate intensity. If now we associate photons with a light-wave (as we know we must), we see that the action of the grating must be represented by saying that the grating distributes the photons among the spectral components in proportion to the intensity of each component. Before passage through the grating, each photon could not be considered as possessing a determined frequency, since each one was bound to a wave involving several frequencies; but after passage through the grating, the photon has a precisely determined frequency, and the probability of each possible value of the frequency is proportional to the intensity of the corresponding Fourier component in the incident wave. The grating is thus a frequency-measuring device which permits us to attribute to the photon a well determined frequency and, consequently, to attribute the value $h\nu$ to its energy and a value $h\nu/c$ to its momentum. One can further say that the spectral composition of the incident wave represents the possible values, after passage through the grating, of the frequency of the photon (that is, of its energy and its momentum) and the respective probabilities of its manifestation after the action of the grating. The possible values are those which figure in the Fourier development of the incident wave, and the probabilities are the squares of the corresponding amplitudes. Applied to a material particle such as an electron, the same correspondence between the values of the energy and the momentum on one hand, and the spectral decomposition of the associated wave on the other, leads to the principle of spectral decomposition as given above.

The special case that we have just examined supplies us with a picture of the role of measurement in Quantum Physics which it seems possible to generalize for all quantities. The measuring apparatus is, finally, always a device that allows us to dissociate the various components of a certain development of the Ψ wave corresponding to the nature of the quantity being measured. It is important to note that the measurement can be accomplished only if the device permits us to separate *in space* the various components of the wave. Thus, in the case of the grating, if we consider the region close to the grating-slits where all the diffracted bundles are superposed, the presence of the photon would not allow us to attribute a frequency and a direction of propagation to it. But both the grating and the incident bundle have finite lateral dimensions. That is why the diffracted bundles always finally separate in space after leaving the grating, and the photon, if it is situated in one of the bundles thus separated in this way, must then possess a well determined frequency and a well determined direction of propagation.

In the analysis of every measuring process, it is, thus, essential to take into account the finite dimensions, not only of the measuring device, but of the wave-trains which are involved in the phenomenon. Every measurement is ultimately made by the localization of a particle in a bundle which the measuring device has extracted from the initial wave.

3. The general formalism of the probabilistic interpretation

Summing up, we have arrived at the following general ideas: Every apparatus permitting the exact measurement of one of the quantities characterizing a particle forces the particle to manifest itself in a state where this quantity has a well determined value; but, prior to the act of measurement, Wave Mechanics, at least if limited to the use of Ψ waves, does not permit the assignment of a well defined value to the quantity; it can only assign to it possible values conditioned by probabilities. In order to find these possible values and probabilities, one has to bring about a certain decomposition of the Ψ wave that is initially associated with the particle—a decomposition determined by the nature of the quantity to be measured and appropriate to the measuring device. In this decomposition, every component corresponds to one of the possible values, and its intensity (the square of the modulus of the amplitude) gives the probability of this possible value. After measurement, the measured quantity has a precise value for the particle; but in general this is not so for some other quantity requiring a different device for its measurement.

It has been possible to represent these circumstances by an elegant and exact formalism of which I will give only a very broad and summary view here.

In this formalism, to every physical quantity there corresponds a linear operator that is Hermitian (or self-adjoint). Now, such an operator A allows us to define (with the help of the equation $A\varphi = \alpha\varphi$) a series of real numbers $\alpha_1, \ldots, \alpha_i, \ldots$ which are its eigen-values, and a series of finite functions $\varphi_1, \ldots, \varphi_i, \ldots$, uniform and continuous in the domain D of the variables considered, which are the eigen functions corresponding to the respective eigen-values. The case may arise, moreover, where several linearly independent eigen functions may correspond to one and the same eigen-value. Those are the cases of "degeneracy" on which I will not dwell.

With the reservation that, in the cases of degeneracy, certain precautions must be taken, it may be said that the eigen functions φ_i form a sequence of orthogonal functions, which means that we have

$$\int_D d\tau\, \varphi_i^* \varphi_j = 0 \qquad (i \neq j) \tag{1}$$

Defined by a linear operator, the φ_i are determined only within a complex multiplicative constant. We can thus impose upon them the condition of being "normalized", that is of obeying the relation

$$\int_D d\tau |\varphi_i|^2 = 1 \tag{2}$$

In the expression for the φ_i there then remains only an arbitrary multiplicative factor $e^{i\alpha}$ of absolute value one.

The eigen functions likewise form a "complete" sequence that allows us to develop a function of the variables of the domain D, such as Ψ, in the form

$$\Psi = \sum_i c_i \varphi_i \tag{3}$$

where the c_i are constant coefficients which generalize the classical Fourier coefficients and are given by

$$c_j = \int_D d\tau \, \varphi_j^* \Psi \tag{4}$$

In this way, a certain decomposition of the Ψ wave is made to correspond to every quantity, or, if one chooses, to every apparatus permitting the measurement of that quantity.

We then admit as fundamental principles corresponding to the general ideas developed above:

1. That every measurement of the quantity necessarily supplies one of the eigen-values α_k.
2. That the probability of a measurement supplying the eigen-value α_k is measured by the square of the modulus $|c_k|^2$ of the coefficient multiplying φ_k in the development of the Ψ by means of φ_k's.

In certain cases, on which I will not dwell here, these statements need to be interpreted to fit the particular situation. It is demonstrable that, applied to the cases of measurement of a coordinate or of a component of momentum, these statements permit us to arrive at the principle of interference and the principle of spectral decomposition.

A very important point in this formalism is the following: It can be shown that the necessary and sufficient conditions for the φ_i to be the simultaneous eigen function of two operators A and B (corresponding to two different measurable quantities) is that A and B "commute", that is, that they produce the same result when applied to a function of the domain D in one order or its reverse. We can then write

$$AB \equiv BA \tag{5}$$

and we say that the commutator AB — BA of the two operators is zero. If that is the case, development (3) will be the same for the two quantities A and B, and the same instrument can measure A and B simultaneously, that is, it will be able to furnish at the same time the values α_i and β_i of the two quantities which correspond to the same φ_i with the same probability $|c_i|^2$.

But, if the operators A and B do not commute, the two developments $\Psi = \sum_i c_i \varphi_i$ and $\Psi = \sum_k d_k \chi_k$ by means of eigen functions of A and B cannot coincide, and so the instruments for measuring A and B will be different. If an instrument, for example, is adapted to the first development, it will permit an exact measurement of A and will be able to furnish one of the exact values α_i with probability $|c_i|^2$ for A; but once this measurement of A has been made, since φ_i does not coincide with any of the χ_k, the development by means of χ_k of the Ψ, after the measurement, will involve several non-zero terms. Thus every measurement will always involve uncertainty of some sort concerning at least one of the quantities A and B.

That is what happens especially for the "canonically conjugate" quantities such as x and p_x to which correspond the operators

$$\mathrm{A} = x \qquad \mathrm{B} = i\hbar \frac{\partial}{\partial x} \tag{6}$$

for which the commutator AB — BA, equal to $-i\hbar$, is not zero. This explains why in connection with these quantities there are inevitably, in every act of measuring, uncertainties Δx and Δp_x such that $\Delta x \Delta p_x \geqq h$.

In this formalism, we designate by the average value $\bar{\mathrm{A}}$ or $\langle \mathrm{A} \rangle$ of the quantity A the mathematical expectation of the value before measurement, that is

$$\bar{\mathrm{A}} = \langle \mathrm{A} \rangle = \sum_i |c_i|^2 \alpha_i \tag{7}$$

and it is easily shown that we have

$$\bar{\mathrm{A}} = \int_{\mathrm{D}} \Psi^* \mathrm{A} \Psi \, d\tau. \tag{8}$$

I will here leave aside important questions, such as that of the primary integrals, and content myself with recalling that one is led to make extensive use of the "matrices" generated by an operator A in the system of the eigen functions Ψ_i of the Hamiltonian operator.

The elements a_{ik} of the matrix A are defined by

$$a_{ik} = \int_{\mathrm{D}} d\tau \, \Psi_i^* \mathrm{A} \Psi_k \tag{9}$$

and obey the rule of non-commuting multiplication

$$(ab)_{ik} = \sum_j a_{ij} b_{jk} \tag{10}$$

The whole formalism that has been summarized here applies, moreover, to the case of systems of particles by replacing ordinary space with configuration space and the particle with the representative point of the system in configuration space.

Chapter VI

VARIOUS ASPECTS OF THE PROBABILISTIC INTERPRETATION OF WAVE MECHANICS

1. The notion of superposition

We have just seen that each function φ_i of an operator A corresponding to a measurable quantity ("observable" in Dirac's sense) describes a state of the system in which the quantity A has the precise value α_i. But in general the Ψ of the system is not reducible to a single φ_i; it is equal to a sum of φ_i's of the form $\sum_i c_i \varphi_i$. We say, then, that the Ψ is a "superposition" of φ_i's. This expression comes from the principle of superposition of very small motions, well known in the classical theory of vibrations.

But here the term "superposition" does not at all have the same meaning as in the classical theories. It is no longer a question of the vibration of a medium which may be obtained by adding up a few elementary vibrations. Rather, we are concerned with the following affirmation: *If the wave-function Ψ of a particle (or of a system) is of the form $\Psi = \sum_i c_i \varphi_i$, and if one seeks to attribute a certain state φ_k to this particle by measuring a quantity* A, *there is a probability equal to $|c_k|^2$ that one will actually be led by this measurement to attribute that particular state to the particle.* Thus, prior to measurement, the particle (or the system) finds itself *potentially* in several states φ_i, each possessing a non-zero probability $|c_i|^2$. That, at least, is what the orthodox probabilistic interpretation says. Now that is an entirely new idea, wholly foreign to classical conceptions, which considers the state of a system as always being characterized by well defined values for all the quantities of the system. This new conception of superposition is one of the most important that the new Mechanics has introduced.

In the classical theory of vibrations, when one considers a vibration represented by an expression of the form $\sum_i c_i a_i e^{2\pi i\left[\nu_i t - \frac{z}{\lambda_i}\right]}$, it means that the magnitude of the vibration is given at every point and at every instant by the addition of plane monochromatic waves, each of whose contributions is measured by the values of the c_i. In Wave Mechanics, the condition $\int_D d\tau\, |\Psi|^2 = 1$—which is imposed on Ψ so that $|\Psi|^2$ is in *absolute value* the probability of localization—no longer allows us to regard Ψ as having a physically determined amplitude. Ψ can therefore no longer represent a vibration having the classical objective meaning.

Let us make this clearer with an example: In Classical Physics, two wave motions $\Psi_1 = c_1 e^{2\pi i\left(\nu t - \frac{z}{\lambda}\right)}$ and $\Psi_2 = c_2 e^{2\pi i\left(\nu t - \frac{z}{\lambda}\right)}$ of the same frequency and direction of propagation give, by superposition, a wave motion $\Psi_1 + \Psi_2$ having an amplitude $c_1 + c_2$. On the other hand, in Wave Mechanics, the states associated with the waves Ψ_1 and Ψ_2, having the expressions just given, must, if we consider them separately, satisfy the conditions of normalization $|c_1| = |c_2| = 1/\sqrt{\mathscr{V}}$ where $\mathscr{V}$ is the volume of the domain D. If we superimpose these states, the Ψ wave becomes $\Psi_1 + \Psi_2$, but with the condition of normalization $|c_1 + c_2| = 1/\sqrt{\mathscr{V}}$ so that we do not have at all the simple addition of amplitudes. This example clearly shows the gulf separating notions of wave-functions in Classical Physics from those in standard Wave Mechanics.

It is scarcely necessary to point out that, if the superposition as defined above is valid for Ψ functions, this is so because these functions are solutions of a *linear* equation of propagation. Since superposition is an indispensable condition of the probabilistic interpretation of the Ψ wave, we are forced to conclude that the Ψ wave obeys a linear equation of propagation. But there will be occasion to ask ourselfves if, by introducing another type of wave, which obeys a non-linear equation of propagation, we might not be able to go beyond the purely probabilistic point of view and thus penetrate to a deeper stratum of physical reality.

2. The equivalence of representations. The theory of transformations

The general ideas allowed by the probabilistic interpretation naturally lead to the notion of the equivalence of all representations of Ψ which, we may say, correspond to various hypotheses — all equally admissible — concerning the measurement that we propose to make.

Let us consider, for example, the developments of Ψ which correspond respectively to a measurement of position in space and a measurement of momentum—measurements which, we know, are mutually exclusive, since no instrument can at the same time give us both an exact localization and the state of momentum. For localization at a point $\mathbf{r}_0(x_0, y_0, z_0)$ in space, it is shown that the eigen-function is the Dirac δ-function

$$\delta(\mathbf{r} - \mathbf{r}_0) = \delta(x - x_0)\,\delta(y - y_0)\,\delta(z - z_0),$$

and, since Ψ may be written

$$\Psi(x, y, z, t) = \int \mathrm{d}x_0\,\mathrm{d}y_0\,\mathrm{d}z_0\,\delta(\mathbf{r} - \mathbf{r}_0)\Psi(x_0, y_0, z_0, t), \qquad (1)$$

we see that the coefficient of the eigen-function $\delta(\mathbf{r} - \mathbf{r}_0)$ is $\Psi(x_0, y_0, z_0, t)$, which is in accord with the principle of localization, since the probability of localization at the point $\mathbf{r}_0$ at time t is $|\Psi(x_0, y_0, z_0, t)|^2$. For the momentum $\mathbf{p}$, the eigen-functions are the plane monochromatic waves $ae^{i(kct-\mathbf{k}\cdot\mathbf{r})}$ where we have put

$$k = \frac{W}{\hbar c}, \qquad \mathbf{k} = \frac{\mathbf{p}}{\hbar} \qquad (2)$$

and where K is a known function of $\mathbf{k}$. Ψ then is developed in the form

$$\Psi = \int \mathrm{d}\mathbf{k}c(\mathbf{k})a_{\mathbf{k}}\mathrm{e}^{i(kct-\mathbf{k}\cdot\mathbf{r})} \quad \mathrm{d}\mathbf{k} = \mathrm{d}k_x\,\mathrm{d}k_y\,\mathrm{d}k_z \qquad (3)$$

and the probability that a measurement of momentum will furnish value $\mathbf{k}$ is given by $|c(\mathbf{k})|^2$.

The equivalence thus appears complete between developments (1) and (3) and all the other developments of the type $\sum_i c_i\varphi_i$ which one might have to anticipate for Ψ when considering other physical quantities. This equivalence serves as the basis for a very elegant mathematical theory due to Dirac and known by the name of the "theory of transformations".

Without going into the development of that theory, let us give serious consideration to its basic idea, limiting ourselves to the case of position and momentum. The basic idea of this theory affirms that, in the state represented by the Ψ under consideration, the particle is potentially present at every point x_0, y_0, z_0 in space with the probability $|\Psi(x_0, y_0, z_0, t)|^2$ and that it also possesses potentially all the momenta $\mathbf{k}$ with the probabilities $|c(\mathbf{k})|^2$. One can then say that the position and momentum both exist, at least in a potential way, prior to the act of measurement to which one of these characteristics of the particle will assign a value.

Does this equivalence suggested by the symmetry of developments (1) and (3) and permitted by the theory of transformations impose itself absolutely? In my opinion it does not—a point that will have great importance in the further exposition of my ideas. In point of fact, what is actually always recorded in an observation of a particle is its position. When, by means of an apparatus, such as a prism or optical grating, we separate the wave components corresponding to the different values of **k**, it is only by a localization of the particle's presence within the spatial extent of these components that we succeed in attributing a momentum to the particle. Moreover, the action of a particle at a point—for example, the local impression made on a photographic plate by the impact of a photon or electron—is a phenomenon that does not require any special apparatus for its observation (other than the purely passive presence of the photographic plate). But the same thing is not at all true of momentum, which for its measurement requires an apparatus that acts on the particle.

Some twenty-five years ago, these considerations led me to believe that the probability $|\Psi|^2$ relative to the localization had, contrary to the basic idea of the theory of transformations, a much more direct significance than the probability $|c(\mathbf{k})|^2$ relative to momentum. The former would in fact express the probability that a particle would be at a certain point in space in the initial state described by Ψ. The latter, on the contrary, would come into existence only after the act of measuring the momentum. Since the measurement performed, along with its result, is still completely unknown, the probability that the value obtained will be **k** would be $|c(\mathbf{k})|^2$. This is the point of view of the causal interpretation to be developed in the second part of this work.

The opposition of these points of view is related to a controversy celebrated in the history of Classical Optics. Certain authors maintained that, when a non-monochromatic wave-train traversed an apparatus of the prism or grating type, the frequencies observed after passage into the instrument were created by the action of the instrument. Other authors said, quite to the contrary, that the frequencies already existed in the original wave-train. From the mathematical point of view, it was the latter group that was right, since the Fourier development of the incident wave is analytically possible and gives rise to the frequencies separated by the prism. But from the physical point of view, within the framework of classical ideas, I think there is no doubt that the former opinion was correct. Indeed, according to classical conceptions, the wave-function represents an objective vibration; it is the time-function, usually a very complex one, representing this objective vibration at every point, which has a physical meaning; and not the Fourier development, which is purely mathematical. In other words, it is the resultant amplitude of the vibration that has a direct physical significance, and not the Fourier components. These components take on a physical meaning only if they are isolated by decomposing the vibration with some instrument for harmonic analysis (analogous to an instrument for measuring momentum in Quantum Mechanics). It is this point of view, most certainly correct for Classical Physics, that we will adopt for the causal interpretation to be set forth presently.

I will cite one more argument which, in Classical Physics, was used to reject the idea that frequencies exist prior to the action of the prism. Let us consider a wave-train of limited dimensions striking a prism. It can be represented by a superposition of plane monochromatic waves that destructively interfere with each other outside of the limits of the wave-train. If these plane waves actually existed in the incident light, since a plane wave has no limits in space and time, the monochromatic components which leave the prism must have existed even before the incident wave-train has reached the prism—which is physically absurd. The wave-train can interact with the prism and split up, as a result of this interaction, into sensibly monochromatic components only at the instant it reaches the prism. This clearly shows that in a wave-train only the overall wave-function has a physical significance, while the Fourier decomposition exists only in the mind so long as the prism has not separated the components of the wave.

Drawn as they are from classical conceptions, the foregoing considerations do not, of course, prove that the idea of absolute equivalence of all the developments of the Ψ function—the idea postulated by the theory of transformations—is incorrect. What they do prove, however, is that we are not categorically forced to accept this equivalence.

3. Wave Mechanics and Quantum Mechanics

In the theory of representations, we consider the equations of motion for the coefficients c_k as being equivalent for any physical quantity. Those equations of motion, for the Dirac constants c_k, have the general form

$$\frac{dc_k}{dt} = \frac{i}{\hbar} \sum_j H_{kj} c_j \tag{4}$$

where H_{kj} is the kj element of the matrix corresponding to the energy. However, when we consider the quantity "position in space", that is, when we adopt what the theory of representations calls "representation q", we note that the equation (4) is nothing more nor less than the equation of the Ψ wave. Indeed, we must then put

$$\Psi(M, t) = \int dP\, \delta(M - P)\Psi(P, t) \tag{5}$$

and equation (4) gives us

$$\frac{\partial}{\partial t}\Psi(P, t) = \frac{i}{\hbar}\int dQ\, H_{PQ}\Psi(Q, t) \tag{6}$$

Now

$$H_{PQ} = \int \delta(M - P)H_M\,\delta(M - Q)dM = H_P\,\delta(P - Q) \tag{7}$$

H_P being the value of the Hamiltonian operator at the point P. There results, then

$$-i\hbar\frac{\partial}{\partial t}\Psi(P, t) = \int dQ\, H_P\Psi(Q, t)\,\delta(P - Q) = H_P\Psi(P, t) \tag{8}$$

which is nothing more nor less than the equation of propagation of the Ψ wave.

So, in the case of the representation q, equation (4) takes on the form of an equation of propagation involving partial derivatives with respect to the spatial coordinates. Representation q thus has a special property: it brings to light a wave aspect that is linked to an equation of propagation. If we consider this fact as basic, and if we attach special importance to this wave-propagation, we will be led to retain the suggestive name of "Wave Mechanics". If, on the other hand, we wish to consider, as does the currently orthodox interpretation, all the representations as equivalent, and to limit ourselves to an abstract formalism without physical imagery, we will prefer the name of "Quantum Mechanics". That is why the choice between these two names for the New Mechanics has more importance than is often believed.

The first point of view is most certainly the one I prefer. Here, for example, is one of the reasons that lead me to believe that the representation q has more physical meaning than the others: Let us consider the problem of determining the stationary waves of an electron in a rectangular enclosure. This problem, as we know, leads us to consider only certain values **p** as possible, defining a certain regular network of points which are "allowed" in the space p_x, p_y, p_z. Now the problem can be stated clearly only for ordinary physical space, for it is only in physical space that we are confronted with the conditions of finite boundaries. It is because the propagation of the wave element is limited by the presence of obstacles—which in this instance are the walls of the enclosure—that quantization seems bound up with the existence of stationary waves. This situation does seem to give a privileged role to the representation q, in short, to physical space.

4. The notion of complementarity (Bohr)

First let us clear up an important point. In elementary treatises on Optics, the simple designation *waves* is generally given to plane monochromatic waves. This arises from the fact that, in practice, ordinary light-wave trains are long enough, albeit limited, to permit us to describe them, along almost their entire extension, as a plane monochromatic wave. A "wave" thus defined has a well defined frequency, wave-length and direction of propagation. Wave Mechanics has this wave correspond to a momentum vector **p**, which points in the direction of the motion and is related to the wave-length by $\lambda = h/p$. The vector **p** thus suffices to define the wave under consideration.

This plane monochromatic wave is homogeneous and does not permit localization of the particle; it is the idealization of the idea of motion without any spatio-temporal localization. On the contrary, the coordinates x, y, z of the particle correspond to the idea of spatial localization at a time t. The canonically conjugate variables p_x, p_y, p_z and x, y, z respectively thus correspond to the wave aspect of the particle which is purely dynamic and without localization, and to the corpuscular aspect with spatio-temporal localization excluding, in a certain sense, the idea of motion. If we then go back to the Heisenberg inequalities, we see that a particle at the atomic level is represented by a plane wave or a localized "granule" only in extreme cases. In general, the plane-wave aspect and the localized-granule aspect both exist, but are both rather hazy, since the associated Ψ wave if formed by a superposition of a certain number of plane monochromatic waves and since, likewise, the localization remains uncertain over a more or less extended region of space.

The uncertainty relations teach us that the more precisely an observation permits us to determine one aspect of the particle, the vaguer becomes its other aspect. This fact makes it possible to explain how Wave Mechanics can simultaneously employ the two conceptions, contradictory in appearance, of the homogeneous plane wave indefinitely extended and of a localized granule. The reason is that these two very different pictures can never flagrantly contradict each other, since one tends to fade when the other becomes more distinct. Bohr has expressed this state of affairs by saying that the plane wave and the localized particle are "complementary aspects" of reality. Every time the behavior of the entity "particle" can be represented by the propagation of a plane monochromatic wave, its particle aspect disappears; and every time this behavior can be represented by the displacement of a particle well localized in space, its wave aspect disappears.

The idea of complementarity, although a bit elusive, is an interesting one. Attempts have been made to apply it in various fields—a procedure that is not always entirely safe. But, from the fact that measuring processes do not permit us to assign a position and a state of motion simultaneously to a particle, are we necessarily obliged to conclude that, in reality, the particle has *neither position nor velocity*?

5. Reduction of the probability packet by measurement

In the interpretation of Wave Mechanics measurement plays a fundamental part. It is measurement that, by supplying us with new information, changes the state of our knowledge about the particle or system under study and suddenly modifies the form of the Ψ function representing this knowledge. If, for example, the measurement is a more or less precise measurement of position, the wave-train representing Ψ prior to measurement will be "reduced" to a less extensive wave-train, perhaps having almost a point-value—if the measurement is *very* precise—whence the name "reduction of the probability packet" which Heisenberg gave to this sudden modification of Ψ. If, on the other hand, the measurement consisted of a determination of the components of momentum, the sudden reduction of the wave-packet would take place in the momentum space, and not in coordinate space.

The reduction of the wave-train gives rise to a new situation which was unforeseeable in advance, since only the probabilities of the various possibilities could be calculated prior to measurement. After a "maximal" experiment, that is, one which supplies the maximum of compatible data, we can construct—with the help of the theory of non-commuting quantities and of the uncertainty relations—a wave function representing our knowledge after the measurement; and we can then follow its evolution in the course of time with the help of the wave equation until such time as we know the result of subsequent measurements that again modify the state of our knowledge and suddenly interrupt the regular evolution of the Ψ wave. The regular evolution of the Ψ wave between two measurements is governed by the wave equation. It is entirely determined by the initial form of Ψ, since the equation of propagation is linear in t. There is, thus, a determinism of the evolution of probabilities between two measurements, but *no* determinism of the sequence of observable facts.

Bohr insisted on the fact that the measurement has the effect of completely wiping out the phase relations between the components of Ψ. Indeed, if the measured quantity A corresponds to the eigen-functions φ_i and if prior to measurement we have $\Psi = \sum_i c_i \varphi_i$, the measurement isolates one of the functions φ_k so that after the measurement $\Psi = \varphi_k$, but the measurement gives us no information whatever on the phase relations of the c_i. If the same measurement were successively made on an infinite number of particles all having the same function $\Psi = \sum_i c_i \varphi_i$ before measurement, the statistical distribution of the values obtained would give the $|c_i|^2$'s, but would still not give us the phases of the c_i's.

The effect of the elimination of the phases by measurement is to make the act of measurement an unbridgeable gap in the evolution of Ψ, both in the direction past-future and in the direction future-past. Now the phase differences between the components in the development of Ψ have a basic importance, for every piece of information about the Ψ function which does not include a knowledge of the phases is radically incomplete. The importance of the phases is brought out very clearly by the study, which is so important in Wave Mechanics, of the interference of probabilities.

6. The interference of probabilities

Let us consider two observable quantities A and B, which we will assume to be non-commuting. The values and eigen-functions of A are α_i and φ_i, those of B are β_i and χ_i. The system of the φ_i and that of the χ_i cannot coincide, since the operators A and B do not commute.

Let us suppose that the initial state is represented by the wave function $\Psi = \sum_i c_i \varphi_i$. Since the χ_i form a complete system, each φ_i can be expressed by the form

$$\varphi_i = \sum_k s_{ik} \chi_k \tag{9}$$

the s_{ik} being the elements of a unitary matrix S. We then have

$$\Psi = \sum_i c_i \varphi_i = \sum_{i,k} c_i s_{ik} \chi_k. \tag{10}$$

If, in the system in the Ψ state we measure quantity A, we find one of the eigen values α_i, the probability of α_j being $|c_j|^2$. After measurement of A, the system will be in the state φ_i, and, in that state, a measurement of B leads to the value β_k with the probability $|s_{ik}|^2$. The total probability of finding the value β_k for B by first measuring A and then B is, then, $\sum_i |c_i|^2 |s_{ik}|^2$.

But let us now suppose that we have measured B directly in the initial Ψ state. Then, according to (10), the probability of finding the value β_k for B is $|\sum_i c_i s_{ik}|^2$. It is entirely different from the preceding probability because it depends on the relative phases of the c_i's, whereas the preceding probability does not. The fact that the probability of the value β_k of B measured directly in the initial state is $|\sum_i c_i s_{ik}|^2$ and not $\sum_i |c_i|^2 |s_{ik}|^2$ may, at first glance, appear contrary to the theorem of composite probabilities, but in reality it is nothing of the sort. The probability $\sum_i |c_i|^2 |s_{ik}|^2$ is indeed the one we should have when the measurement is taken first of A and then of B, since it is equal to the sum of the products of the probability for obtaining first a value α_i of A by the probability of obtaining β_k for B *when we know that* α_i *for* A *has already been obtained.* The theorem of composite probabilities thus remains intact, but there is no reason why the probability $\sum_i |c_i|^2 |s_{ik}|^2$ should be equal to the probability of obtaining the value β_k of B directly from a measurement of B in the initial state. What gives rise to a certain confusion in this question is that in mathematical statistics the measurement of a probability variable (always of a macroscopic nature in ordinary statistics)—a measurement that statisticians usually call a "trial"—in no way modifies the probabilities relative to other probability variables. Thus, if we wish to set up statistics on the height and chest-circumference of a batch of conscripts, these two quantities are measured on all the conscripts, and it is taken for granted that the height measurement can in no way effect chest-circumference and vice-versa. If x stand for height and y for chest-circumference, we will have

$$\text{Prob}\,(x_k) = \sum_i \text{Prob}\,(y_i)\,\text{P}_{y_i}(x_k) \tag{11}$$

where $\text{P}_{y_i}(x_k)$ is the probability of the height x_k for a conscript who has a chest-circumference y_i, and there is no need to state whether measurement x was made before or after measurement y.

But these hypotheses, unquestionably valid in the macroscopic field, are not necessarily so at the microphysical level. At the microphysical level the situation is such that, because of the existence of the quantum of action, the measurement of a probability variable modifies the probability for the other quantities. The probability of B is not the same prior to and after the measuring of A. As we have seen, the probability of a value of B, if we start by measuring A, is correctly given by the theorem of composite probabilities, but it is not equal to the same value of B measured directly in the initial state.

I emphasized these circumstances in an article in the *Revue Scientifique* (1948, p. 259) by giving some familiar examples, and I sought to make clear in what way the schema of probabilities in Wave Mechanics differs from the usual schema of the statisticians. In the usual schema, one defines the densities of probabilities $\rho_X(x)$ and $\rho_Y(y)$ relative to two probability variables X and Y; $\rho_X(x)$ is the probability of a value of X being between x and $x + dx$, and similarly for $\rho_Y(y)$. It is likewise assumed that there exists a density of probability $\rho(x, y)$ corresponding to the possibility of obtaining *in one and the same trial* values x and y for X and Y. Also defined is the probability of Y conditioned by X, $\rho_Y^{(X)}(x, y)$, which corresponds to the probability of obtaining the value y for Y when we know that X has the value x; and we define in the same way the probability $\rho_X^{(Y)}(x, y)$ of X conditioned by Y. Between these five quantities we have the relations

$$\begin{aligned} \rho_X(x) &= \int dy\, \rho(x, y) \qquad & \rho_Y(y) &= \int dx\, \rho(x, y) \\ \rho_X^{(Y)}(x, y) &= \frac{\rho(x, y)}{\rho_Y(y)} \qquad & \rho_Y^{(X)}(x, y) &= \frac{\rho(x, y)}{\rho_X(x)} \end{aligned} \tag{12}$$

where the integrals must be replaced by sums in the case of discontinuous probabilities. From them we obtain

$$\rho_X(x) = \int dy\, \rho_Y(y)\, \rho_X^{(Y)}(x, y) \qquad \rho_Y(y) = \int dx\, \rho_Y^{(X)}(x, y)\, \rho_X(x). \tag{13}$$

Now, in Quantum Mechanics, if we consider two canonically conjugate quantities X and Y, for example $X = x$ and $Y = p_x$, we can define $\rho_X(x)$ and $\rho_Y(y)$, but one can no longer define $\rho(x, y)$, since it is impossible to obtain the value of the canonically conjugate quantities X and Y simultaneously. The quantities $\rho_X^{(Y)}(x, y)$ and $\rho_Y^{(X)}(x, y)$ can still be defined, but we no longer have relations (13) since they resulted, in the classical schema, in relations (12) which here no longer have meaning in view of the fact that $\rho(x, y)$ no longer exists.

If we then admit that, in a state Ψ, all the quantities have distributions of probabilities defined for the possible results of measurements made on the systems in that state, it is impossible to maintain the classical statistical schema with a $\rho(x, y)$ *and* relations (12) and (13). Any attempt made in that direction is doomed to failure within the framework of the standard interpretation of Wave Mechanics.

But is one obliged to admit that all the distributions of probabilities defined by the usual statistical interpretation of Wave Mechanics already exist in the initial Ψ state? As we have seen, such does not seem to be the case. One can readily admit, prior to the knowledge of momentum measurement, that the probability of localization $|\Psi|^2$ exists in the initial state, while the probability $|c(\mathbf{k})|^2$ of a value of momentum does not exist until after making the measurement. In the initial state Ψ, the momentum might very well possess a precise value (not determinable by a measurement which has as a consequence the modification of that value), and the probability of that value, different from $|c(\mathbf{k})|^2$, would permit us to restore the usual statistical schema for the initial state. We will see that this is exactly what is found in the causal interpretation to be developed further on.

The impossibility of maintaining the usual statistical schema in the now-current interpretation seems to arise, thus, from the fact that in it distributions of probabilities are compared which are not simultaneously valid—certain ones being valid in the initial state prior to measurements, the others in the final state after measurement when the result of that measurement is not previously known. It is this possibility that was not taken into account by von Neumann when he concluded that it was impossible to reestablish classical conceptions by the introduction of hidden variables.[1]

[1] As might be suspected from the description of the theory of the Double Solution as *causal*, the theory is able to surmount the difficulties of the usual statistical schema found in a purely probabilistic formulation of particle physics mentioned in this section (eq. 12 et seq.). For a description of the reestablishment of the usual statistical schema, see ref. [17] pp. 88–93.

7. Von Neumann's theorem

In his important work *Mathematical Foundations of Quantum Mechanics*,[2] J. von Neumann has given a most rigorous exposition of the probabilistic interpretation of Wave Mechanics. In particular, he was led to study the theory of measurement very closely. One finds this theory taken up again and explained with exceptional clarity in a fascicule of *Actualités scientifiques et industrielles* (1939, No. 775), by London and Bauer under the title "*La Théorie de l'observation en Mécanique quantique*".[3]

One of the great merits of von Neumann's presentation is the way it clearly distinguishes between the "pure" and the "mixed" cases. One has a pure case when the state of a system is represented by a Ψ function, with the distributions of probabilities of the various quantities being given by the squares of the moduli $|c_k|^2$ of the coefficients in the development of the Ψ according to the eigen-functions of the quantity; these distributions of probabilities are characterized by the interference of probabilities, and, as we have seen, depart from the schema of probability distributions as found in ordinary statistics. On the other hand, one has a mixture when the Ψ function of the system is not precisely known and when one can merely attribute various functions of the order $\Psi^{(k)}$ to it, with probabilities p_k (such that $\sum_k p_k = 1$). Here the coefficients of probability p_k are defined in an altogether classical way. To characterize such a system from the statistical point of view, von Neumann has defined an Hermitian "statistical matrix" whose trace is equal to unity; and to do this, von Neumann utilized the notion of a "projection". For a pure case, the statistical matrix, now called an "elementary statistical matrix", enjoys the property of being idempotent, that is, $\mathrm{P}^n = \mathrm{P}$, no matter what n is. For a mixture, on the other hand, the statistical matrix is not idempotent. The necessary and sufficient condition for the statistical matrix of a system to be idempotent is that the system shall be a pure case. In this way we have a criterion for distinguishing pure cases from mixtures. By carefully analyzing the notion of measurement, von Neumann has shown that the effect of measurement was to transform a pure case into a mixed one, which is tantamount to saying that the measuring device has the effect of isolating wave-trains corresponding to the various components of Ψ for the quantity under consideration,

[2] Princeton University Press (1955).

[3] Hermann, Paris.

with only one of these wave-trains corresponding to a physically realized hypothesis concerning measurement.

It was in the course of this investigation that von Neumann thought he was able to demonstrate the impossibility of explaining the probability distributions of Wave Mechanics by introducing "hidden variables". This demonstration seems to eliminate once and for all the possibility of ever going back to a causal and objective theory of microscopic phenomena.

Without retracing all the steps of the reasoning, let us indicate its general trend: Von Neumann first demonstrated the following theorem: *It is impossible to represent a pure case in the form of a mixture.* In other words, a pure case is never reducible to a sum of pure cases.

Having established this point, von Neumann makes the following remark: If it were possible to obtain a classical interpretation of the probability distributions in Wave Mechanics by the introduction of hidden variables (the way, in Classical Physics, the kinetic theory supplied them for the laws of gases), the knowledge of the exact values of the hidden parameters would, in principle, permit us to obtain a state "without dispersion", that is, a state where for every quantity A, the dispersion $\sigma_A = \sqrt{\overline{(A - \bar{A})^2}}$ would be zero. One would obtain the statistical properties of the system by considering mixtures of these states without dispersion—as the statistical theories of Classical Physics do. In short, for a statistical theory to be reduced to a deterministic schema with hidden parameters, we must be able to reduce the statistical distributions of this theory to mixtures of elementary states that are not decomposable and that have no dispersion. Now von Neumann shows that this is not the case for Wave Mechanics, which, consequently, cannot by any means whatsoever be reduced to a deterministic schema of hidden parameters.

The reasoning rests essentially on the following theorem: *The states encountered in Wave Mechanics can never be devoid of dispersion.* Von Neumann has justified this statement by showing that in Wave Mechanics no acceptable statistical matrix P exists corresponding to an absence of dispersion for all quantities. Moreover, this result can be easily foreseen by noting that even in the pure case (a system having a well determined Ψ function), the dispersions σ_x and σ_{p_x} of two canonically conjugate quantities cannot be simultaneously zero, by reason of the dispersion theorem expressed in the inequality $\sigma_x \sigma_{p_x} \geqq \hbar/2$.

We therefore cannot reduce the probability distributions of Wave

Mechanics to mixtures of undecomposable states without dispersion. Many undecomposable states exist: the pure cases; but they are never without dispersion. Von Neumann's conclusion can thus be drawn solely from study of the pure states, but the general analysis that he has given allows us to make a more precise comparison with the probabilistic theories involving the hidden parameters of Classical Physics.

The beautiful mathematical neatness of von Neumann's deduction might lead one to the conviction that any return to the causal and objective conceptions of Classical Physics was henceforth impossible in Microphysics. It might be objected that the demonstration was based on the postulate that the distributions of probabilities allowed by Mechanics have a general validity; but one could reply that actual experience furnishes complete confirmation of that postulate. One might also say that von Neumann's demonstration did not add greatly to what was already known, since the conclusion is already implied in the uncertainty relations; but this observation in no way diminishes the soundness of his conclusion.

But, as we will see, there exists at least one theory—the causal theory to be studied later—that permits us to obtain the probability distributions of Wave Mechanics and that is at the same time a deterministic theory involving hidden parameters. It may be that this theory is not physically accurate, but it does exist; and its very existence is already a contradiction of von Neumann's theorem. How is that possible?

The examination of this question led me—and on this point I am in agreement with David Bohm—to think that von Neumann's demonstration implies a hypothesis that is not absolutely unavoidable, and that is not substantiated in the causal theory in question. This hypothesis is that when a system is in a state Ψ the distributions of probabilities defined by Wave Mechanics are valid *prior to* any act of measurement. Now, if these laws of probability are stated in the way they should be (for example, $|c_k|^2$ is the probability that an exact measurement of the quantity A will furnish the value α_k), it is immediately clear that they are valid only after the measurement has been performed and prior to the knowledge of the result. For two non-commuting quantities, probability laws ought never to come into play simultaneously, since they can become valid only after measuring operations that are incompatible with each other. It is possible that a certain distribution of probabilities may already be valid in the initial state and be merely confirmed by measurement—that is the case, in the causal theory, with the distribution of $|\Psi|^2$ for localization. But, in general, the distribution of probabilities will be the result of the measuring operation and may follow an unknown distribution of probabilities, perhaps even an unobservable one, that exists prior to measurement. Such is the case, in the causal theory, of the probability $|c(\mathbf{k})|^2$ for momentum.

These observations now seem to me to cast doubt on the validity of the implicit postulate on which von Neumann's demonstration is based, and, as a result, to destroy the penetrating force of his arguments.

Chapter VII

OBJECTIONS TO THE PURELY PROBABILISTIC INTERPRETATION OF WAVE MECHANICS

1. Consequences of the disappearance of the trajectory concept

In the present purely probabilistic interpretation of Wave Mechanics the concept of a trajectory has disappeared, at least whenever one leaves the domain in which Geometrical Optics is valid for the propagation of the Ψ wave. When that approximation is valid, we can retain the notion of a trajectory and consider almost point-like wave-trains describing ray-trajectories; but the moment, for example, that interference and diffraction phenomena occur, the notion of a ray and, consequently of a trajectory, becomes useless. The particle in physical space (or the representative point in configuration space) can be localized by measurement only *at widely separated intervals*, and no trajectory could be assigned to it between the localizations. From this situation important differences would result in regard to the very concept of probability in Classical Physics and in Microphysics, as will be explained.

Let us consider, within the framework of Classical Physics, all the possible motions corresponding to a single Jacobi function S. Jacobi's theory tells us to consider all the trajectories involved as being rays of a wave propagation whose surfaces S = const. are the wave-surfaces. If we are confronted with an infinite number of particles describing all the possible trajectories of the class under consideration, we can imagine that the particles are distributed throughout a cloud with a spatial density $\rho = |\Psi|^2$, Ψ being the wave defined by Jacobi's theory. Indeed, ρ is unchanged in the course of time. We can then also say that $|\Psi|^2$ measures the probability that a given particle will be found at a given point at a given instant. Here probability enters in a wholly classical way as a result of our ignorance of both the trajectory actually described by the particle under consideration and of the particle's position on that trajectory. In principle, the equations of Dynamics would permit us to calculate the trajectory actually described, as well as the motion along this trajectory, if we know the initial position and the initial velocity of the particle. But if some of these data are missing, we will know only the *possible* trajectories, and there will then be only a probability, and not a certainty, of finding the particle at point M at the instant t. If an observation does permit us to discover the presence of a particle at point M at the instant t, we will know that the trajectory described passes through M, and thereafter we can be sure that there will be no chance of discovering the presence of the particle anywhere but along that trajectory. The probability of presence, which was different from zero in an extended region of space, merely expressed our ignorance of the trajectory actually described. It loses all meaning as soon as we know the trajectory. Such is the point of view of Classical Physics—a point of view in keeping with the traditional intuitive concepts of Science. Classical Physics admitted, in particular, the determinism of motions, and probability entered the picture only as a result of our ignorance of the data essential to the perception of this determinism. In this way there was perfect agreement with the conception of Probability as given by all the great masters of Classical Science from Laplace through Henri Poincaré.

The point of view of the present probabilistic interpretation of Wave Mechanics is entirely different. According to this interpretation the conception of a trajectory is only a first approximation, valid only when Geometrical Optics is applicable to the propagation of the Ψ wave. The moment this is no longer the case, and especially when there is interference or diffraction of the Ψ wave, the notion of a trajectory becomes impossible to use, and one is obliged to speak only of successive localizations of the particle in space (or of the representative point in configuration space) which result from observations involving a position measurement.

Probability, then, must be introduced with an entirely new character. It is no longer the expression of our ignorance of a trajectory followed by the particle, since there is no trajectory. Von Neumann even seemed to have demonstrated in his famous theorem that the introduction of probabilities into Quantum Physics could not in any way arise from our ignorance of certain hidden parameters that eluded us. That being the case, there is no longer any determinism; nothing any longer permits us to make exact predictions—save in very exceptional cases—concerning the precise result of a measurement. We can only assign a probability to each possible result of a measurement. Probability would intervene without any ignorance on our part of a hidden situation.

Contrary to the affirmations of all the scientists of the pre-quantum period, probability would exist "in a pure state", without being the result of a determinism that eludes us. This is a very interesting new conception—but one that also gives rise to difficulties.

The description of the microscopic world, tied uniquely to the knowledge of a function, the Ψ wave—which is nothing more than the representation of a probability, and dependent upon the user's knowledge —takes on a subjective character, and the objective character of physical reality is called into question in a rather singular way. We will see especially what strange consequences arise as a result of abandoning the concept of a trajectory.[1]

2. Einstein's objection at the 1927 Solvay Congress

At the Solvay Congress of October 1927, Einstein raised a very striking objection to the purely probabilistic interpretation of Wave Mechanics.

He considered a particle that impinges normally on a plane screen in which there is a small hole; a photographic film having the shape of a hemisphere with large radius is placed behind the screen.

If the hole has small enough dimensions, the Ψ wave associated with the particle will be diffracted as it goes through the hole and spread out on the hemispherical film, for the hole will play the role of a tiny point-source placed at the center. If, at an instant t, a photographic impression reveals the presence of the particle at a point A on the film, the interpretation of this fact will vary greatly, depending on whether we reason according to classical ideas or according to the new conceptions.

Following classical ideas, we must say that the particle going through the aperture necessarily has a "trajectory". This trajectory, represented in Fig. 3 by a broken line, will of necessity strike the film at one of the points on it; but so long as we have not yet discovered the presence of the particle at a point on the film, we will not know the trajectory actually followed by it, and for that reason we will attribute to the particle's presence at every point on the film a non-zero probability (equal to $|\Psi|^2$). As soon as the particle's presence is discovered at A, we know the trajectory and the probability of finding the particle at

[1] For further remarks on difficulties arising from the purely probabilistic approach, see ref. [17], pp. 84—93.

another point B on the film at once becomes zero. All that is perfectly clear.

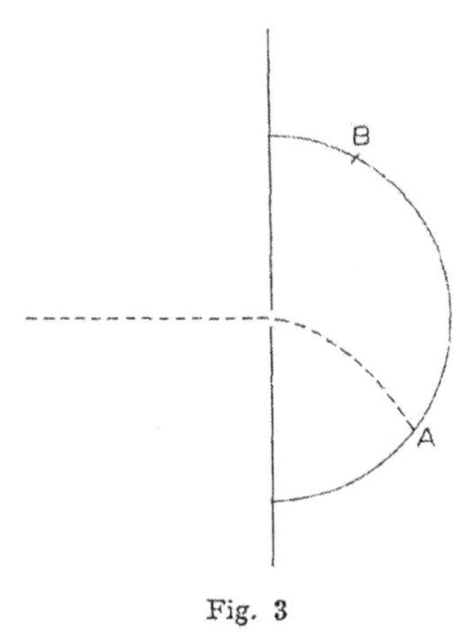

Fig. 3

But, according to the new concepts, we are obliged to admit that there is no definable trajectory, since there is diffraction to the right of the screen. So long as localization at A has not taken place, the particle must be considered present, in a potential state, over the whole surface of the film, with a probability $|\Psi|^2$. As soon as the particle has shown its presence at A, the probability of finding it at any other point on the film instantly becomes zero, since, by hypothesis, there is only one particle associated with the Ψ wave. The interpretation of this fact, which is perfectly simple when one can admit the existence of a trajectory, now becomes most mysterious. It is, in fact, with our classical ideas of space and time (and even with relativistic ideas of space-time), impossible to understand how the fact of observing an effect localized at A can instantly render impossible an analogous effect at every other point B on the film, even very distant from A, if one does not admit that the particle is localized in space at every instant and describes in the course of time a well defined trajectory, perhaps unknown to us.

If then, going along with the New Mechanics, we abandon the notion of a trajectory, it is necessary to admit that the particle—even though it is an indivisible unit and well localizable at a given instant—is not constantly localized in space and in time. It is, so to speak, "virtually" present in the whole wave-train, and one will have to say with Bohr that "particles are individual entities defined in an indefinite way over extended regions of space-time" and that their behavior "transcends" the framework of space-time—a kind of philosophical language that is, perhaps, rather dangerous to introduce into Theoretical Physics.

In Einstein's example, the particle would in a sense be spread out in a virtual state in the space beyond the screen. At the moment an effect localized at A takes place, the particle would, so to speak, condense at that point in order to produce an observable phenomenon. Now, as Einstein has emphasized, no mechanism compatible with the older ideas, even when relativistic, about space and about time can account for this sudden contraction of the particle—a contraction that would result in the instantaneous action of an event taking place at A, on whatever may take place at the distant point B. The present interpretation of Wave Mechanics would thus oblige us to regard the standard notions of space and time as totally inaccurate, not only on the microphysical level (which would still be acceptable), but even on the macroscopic level, since points A and B can be considered very far from each other on the film.

It is thus legitimate to consider Einstein's example as a very serious objection to the present interpretation of Wave Mechanics—one that has never been clearly answered.

3. The example of Einstein, Podolsky and Rosen

There have been lively and interesting discussions, in which very eminent scientists have participated, on the subject of "correlated" systems, that is, of systems which, once having been in interaction find themselves subsequently separated from each other, but in states with probabilities that are no longer independent. These debates were set off by a paper by Einstein, Podolsky and Rosen,[2] which was commented on by Schrödinger.[3] Bohr replied to the paper in an article in *The Physical Review*,[4] and still further remarks were presented by Furry.[5]

Einstein, Podolsky and Rosen had first considered a two-particle system, the state of which was represented by the wave function (where d is a non-zero constant)

[2] *Phys. Rev.*, 47 (1935) 777.
[3] *Naturwissenschaften*, 23 (1935) 787, 823 and 844.
[4] *Phys. Rev.*, 48 (1935) 696.
[5] *Phys. Rev.*, 49 (1936) 393.

$$\begin{aligned}\Psi &= \int da \int db\delta(x_1 - a)\delta(x_2 - b - d)\delta(a - b) = \int da\delta(x_1 - a)\delta(x_2 - a - d) \\ &= \int dk_1 \int dk_2 \delta(k_1 + k_2) e^{2\pi i(k_1 x_1 + k_2 x_2)} e^{-2\pi i k_2 d} \qquad (1) \\ &= \int dk_1 e^{2\pi i k_1 (x_1 - x_2 + d)} = \delta(x_1 - x_2 + d)\end{aligned}$$

The transition from the expression in the first line to that in the second is effected by recalling that we have for the Dirac δ-function the symbolical expression

$$\delta(x) = \int_{-\infty}^{\infty} dk e^{2\pi i k x} \qquad (2)$$

Form (1) of the Ψ shows that the measurement of k_2 always entails $k_1 = -k_2$ and that the measurement of x_2 always entails $x_1 = x_2 - d$. In other words, the quantities $k_1 + k_2$ and $x_2 - x_1$ have, respectively, the values 0 and d, which is possible, since the corresponding operators commute (although k_1 and x_1 on one hand, and k_2 and x_2 on the other, do not commute).

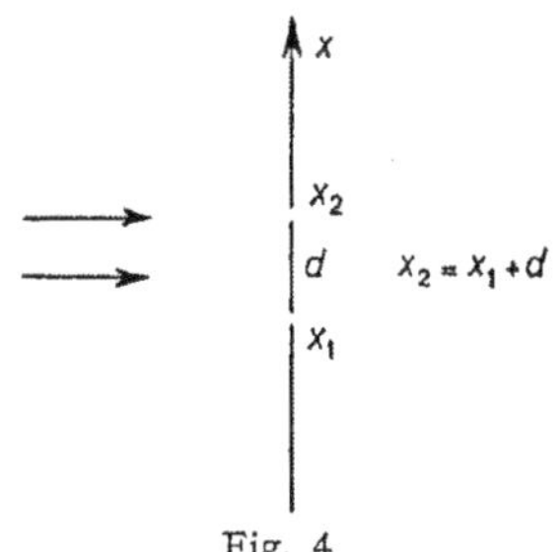

Fig. 4

Form (1) of Ψ can be interpreted physically by considering a plane screen with two very small parallel slits in it, and which is struck at normal incidence by plane monochromatic waves initially associated with the two particles.

If, in the initial state, we know precisely what the motion of the screen along the x-axis (its p_x) is, the position of the screen along this axis, and consequently the abscissa of the first slit also, must remain unknown according to the Heisenberg uncertainty relations, since all the values of x_1 are equally probable. The value of the Ψ of the system immediately to the right of the screen is then given by $\int da\, \delta(x_1 - a)\delta(x_2 - a - d)$ which expresses the simultaneous presence of particle 1 in the first slit and particle 2 in the second slit, with the position of the double-slit system being undetermined. If the momentum of the screen does not vary when the particles pass through the slits, one should have $k_1 + k_2 = 0$, in agreement with the third expression of (1) for the Ψ.

The point to which Einstein and his collaborators drew attention is this: Since the initial state is given by (1), we are free to measure either x_2 or k_2—which will lead us either to attribute the value $x_1 = x_2 - d$ to the x-coordinate of the first particle, or to attribute the value $k_1 = - k_2$ to the conjugate component of its momentum. Since neither of these measurements affects the first particle, we can attribute to it either a position or a momentum along the x-axis *without in any way acting on the particle.* And we can do this, if the orthodox interpretation is correct, even though this position and this momentum cannot have precise values at one and the same time. Einstein, Podolsky and Rosen concluded from this that particle 2 must have, prior to the measurement made on 2—which in no way effects 1—a well determined position and momentum; and, consequently, the description of reality by means of the Ψ wave is, if not inaccurate, at least incomplete.

Unfortunately, Einstein, Podolsky and Rosen's example does not seem a happily-chosen one because, in the state defined by form (1) of the Ψ, the two particles cannot be considered "separated" in space, and because both of them interact with the same experimental apparatus—namely, the screen with the double slits.

That is what made it possible for Bohr to reply to his contradictors in a way that seems satisfactory. He noted that, in view of the assumed apparatus, the two possible measurements (those of position and momentum) correspond to different experimental arrangements. The measurement of the positions presupposes that we fix the position of the screen in relation to the macroscopic set-up that allows us to establish our spatial coordinates. Then the first slit will have an exact abscissa $x_1 = x_0$, and the second an abscissa $x_2 = x_0 + d$. But knowledge of the momenta will be entirely lost because, the slits being rigidly fastened to the set-up, the momentum that the screen might receive from the particles will be lost in the supporting structure. Inversely, if one wishes to measure the momenta, it will be necessary to measure the initial impulses of the screen—which presupposes that we allow the screen to be mobile and, consequently, it is impossible to know the abscissae of the slits exactly. In this case, the variations of the momentum of the screen along the x-axis equal to $K_0 - K_1$ being known, the measurement of k_2 will give the value $k_1 = - k_2 + K_0 - K_1$ for k_1. Bohr's conclusion is, thus, the following: In order to consider the problem, it is necessary to indicate precisely all the details of the experimental apparatus; for, just as soon as the actual measurement is under way, one must have a structure suited to the two measurements that one desires to make.

Bohr's reasoning, on occasion rather nebulous, contains a number of questionable assertions. For example: The measurement of a particle's impulse is always made by communicating that impulse to a macroscopic body to which the conceptions of Classical Physics are applicable. It seems to us, on the contrary, that such is never the way the impulse of a particle is measured, but rather by deducing it from the observed localization of another particle with a possible application of the conservation of impulse. Nevertheless, it must be granted that, by reason of Einstein, Podolsky and Rosen's none-too-happy choice of examples, Bohr has been able to dispose of the objection in the form in which it was presented.

4. An objection relative to correlated systems (Schrödinger)

The weakness of Einstein, Podolsky and Rosen's example seems to me to arise from the fact that it does not explicitly bring in the spatial limitation of all wave-trains. Physically, the wave associated with a particle cannot be a plane monochromatic wave indefinitely extended in space and in time; we are always of necessity concerned with a spatially limited wave-train. If the waves associated with two particles were strictly plane and monochromatic, one could not speak of a collision between them, since the present interpretation assumes that in that case they would be potentially present at every point in space, and hence constantly and indefinitely in a state of collision. So physically, it is always necessary to take into account the limited extension of the wave-trains—an all-important point to which we will constantly be referring.

Moreover, we are now going to present the objection relative to correlated systems which explicitly brings in the limited dimensions of the wave-trains. In *that* form, which is very close to the one adopted by Schrödinger in the articles cited above, the objection strikes me as very difficult to overcome.

Let us consider two groups of almost monochromatic waves associated with two particles 1 and 2, and let us assume that they are about to collide with each other.

When they have reached the shaded region R, the two particles interact and their waves are superposed. In order to predict what is going to happen, one must consider the Ψ wave of the system in configuration space. Wave Mechanics tells us that the collision can give rise to a whole series of possible final motions, all compatible with the conservation of the energy and of the components of impulse. Now either the wave-train of particle 1 will finally describe the trajectory 1′ while the wave-train of particle 2 describes 2′; or else the wave-train of 1 will finally describe the trajectory 1″ while the wave-train of 2 describes trajectory 2″, and so on. The final wave-function of the system of the two particles in configuration space will be a superposition of the products of the wave-function for 1′ and the wave-function for 2′, of the function for 1″ and function for 2″, etc.—the coefficients of the superposition giving us by the square of their modulus the probabilities of the various correlated states 1′ — 2′, 1″ — 2″, etc.

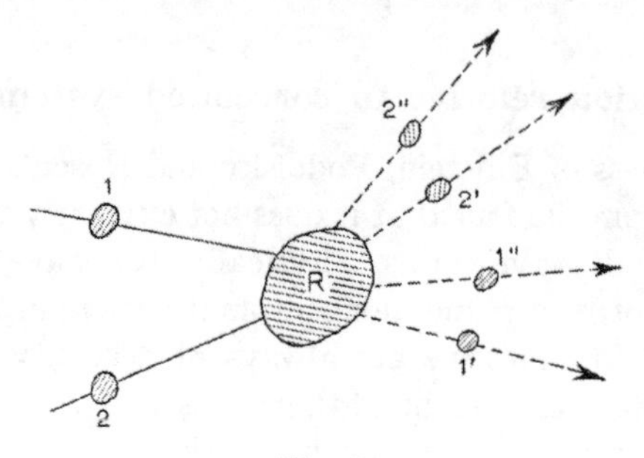

Fig. 5

Let us now suppose that we place a counter on trajectory 1′ capable of revealing the arrival of particle 1. If this counter records the arrival, we will know that particle 2 is in wave-train 2′. This is easily understandable if the particles occupy a position in physical space at every instant, for then we can say that the trajectories of particles 1 and 2 are correlated in such a way that, if the first particle follows trajectory 1′ after the collision, the second will follow the trajectory 2′, etc. The operation of the counter will simply have revealed a fact that *already* existed—namely, that the first particle followed trajectory 1′. The Ψ wave of the system will be reduced to the packets for 1′ — 2′, for the other packets disappear instantly when the operation of the counter is perceived; and this is quite explicable, since the Ψ wave is only a representation of probability, a subjective element that is suddenly modified by acquisition of new information.

But the point of view we have just adopted amounts to admitting that the Ψ wave is not a complete representation of reality, since it would be necessary to give the particle's position also, in order to have a complete picture—that is, it would be necessary to give the values of those "hidden parameters", the coördinates.

The point of view at present regarded as orthodox affirms, on the contrary, that the description of the system by the function is a complete description and that there is no permanent localization of particles. After the collision, particle 1 is potentially present in the wave-trains 1′, 1″, etc., while particle 2 is potentially present in the wave-trains 2′, 2″, etc. When the counter placed on trajectory 1′ records, particle 2 would be instantly thrown into the single wave-train 2′, although 2′ may be situated as far away from the counter that has just recorded as one likes. As Schrödinger said, "That would be magic."

And here we again come across the contradiction that exists, even on the macroscopic level, between the purely probabilistic interpretation of Wave Mechanics and the notions of space and time—a contradiction that was pointed out by Einstein as far back as the Solvay Congress of 1927.

The exponents of the purely probabilistic interpretation of Wave Mechanics often express themselves in the following terms: *When a system happens to be in a state in which quantity* A *does not have a well determined value, but instead a whole series of possible values, and if we take an exact measurement of* A, *we cause the system to pass into a new state in which* A *has a well determined value*; *in this way, by the very act of measuring* A, *we produce an* uncontrollable *perturbation on the system—a perturbation which makes it impossible for us to know the values of quantities that do not commute with* A. This fact is then explained by pointing out that, in order to measure A, we necessarily exert an effect on the system, an effect which the existence of Planck's constant does not allow us to reduce indefinitely. This explanation, which is no doubt accurate in certain cases, becomes inadmissable for the case of correlated systems that we have just studied. It is, indeed, inconceivable that the counter placed on trajectory 1′ and acting on particle 1 could exert any action whatsoever on particle 2.

To present the difficulty in a slightly different but very striking case, let us consider the emission of a quantum of radiation by a hydrogen atom. To simplify things, we will assume that the atom has a series of stationary states K, L, M, whose fine structure will be neglected. In general, in the initial state of the atom, its Ψ wave will have the form

$$\Psi_i = c_K \Psi_K + c_L \Psi_L + c_M \Psi_M + \dots \tag{3}$$

with the square of the moduli of the c's giving the probability of the various states of energy in this initial state. At a certain possibly very large distance from the atom we place an apparatus capable of revealing the arrival of a photon and even of giving us the photon's frequency, as is the case of the photoelectric cell. If at a given moment this apparatus indicates the arrival of a photon having the frequency of the first line of the Balmer series (corresponding in Bohr's schema to the transition $M \to L$), we will be forced to say, according to the current interpretation, that the atom, initially distributed in a virtual state among the energies E_K, E_L, E_M, ... with probabilities $|c_K|^2$, $|c_L|^2$, $|c_M|^2$, ... has passed into the energy state E_L with emission of the line of frequency $\nu = (E_M - E_L)/h$, and the final state of the atom is represented by

$$\Psi_f = \Psi_L. \tag{4}$$

This would be easily understandable if we could say that the emission process was accomplished by a photon with frequency ν being emitted from the atom—which atom was left in the final state Ψ_L. Recording the arrival of the photon would then merely have told us the sequence of events. This information, by modifying our knowledge concerning the state of the atom, would oblige us to modify the Ψ wave that symbolizes this knowledge.

But the interpretation accepted at the present time is entirely different. We have to say that, so long as the photon-detector has not operated, the atom is in the state Ψ_i. The operation of the detector then instantaneously forces the atom into the state Ψ_f, and this notwithstanding the fact that the distance from the atom to the photon-detector may be very great. Such an interpretation really seems inadmissable, unless the Ψ is a purely subjective element representing the state of knowledge of the physicist employing it. But in that case one can hardly claim that the function furnishes a true description of the phenomenon "emission of a photon by an atom".[6]

[6] The remarks in this section are considerably expanded in Chapter VII of de Broglie, *La théorie de la mesure* ... [17], pp. 98–112.

5. Some other objections by Einstein

In 1949, on the occasion of Einstein's seventieth birthday, there appeared in the United States an anniversary book dedicated to the founder of the Theory of Relativity; and to it, scientists from every country contributed articles.[7] Some of the great quantum physicists such as Born, Pauli, Heitler, etc., expressed in their articles, sometimes in rather sharp terms, their disappointment at seeing Einstein persist in a negative attitude towards the purely probabilistic approach to Wave Mechanics.

The most interesting of the studies included in the book is unquestionably the one due to Bohr, in which the illustrious Danish scientist, after analyzing the Quantum Theory, its early development, and the sudden flowering of Wave and Quantum Mechanics, goes on to summarize the gist of his discussions with Einstein during the period following the Solvay Congress, concerning the interpretation of this Mechanics.

In his reply, which is printed at the end of the volume, Einstein persists in denying that the Ψ function of Wave Mechanics can give, by itself, a complete description of reality. According to him it cannot describe an individual system, but only a statistical aspect of a collection of identical systems. Einstein fully recognizes that the present formalism of Wave Mechanics perfectly describes observable phenomena; but, he says "I am, in fact, firmly convinced that the essentially statistical character of contemporary quantum theory is solely to be ascribed to the fact that this [theory] operates with an incomplete description of physical systems." (p. 666)

As an example, Einstein studies the current theory of α disintegration of a radioactive body. This phenomenon is represented by describing the Ψ wave of the α particle escaping from the nucleus in the form of a divergent spherical wave that progressively filters through the "barrier of potential" surrounding the nucleus. This picture, says Einstein, is perfect, providing one only intends to study the statistical properties of a collection of radioactive nuclei; but it cannot give a really complete description of any one of these nuclei because it cannot accurately indicate the *time* of the disintegration, and because one must obviously assume that every nucleus disintegrates at a well defined time. Einstein

[7] *Albert Einstein: Philosopher-Scientist*, Edited by PAUL ARTHUR SCHILPP, The Library of Living Philosophers, Inc. Evanston, Ill. 1949.

then goes on to present the answer that a partisan of the present interpretation of Quantum Physics would undoubtedly make to him. This answer would consist essentially of pointing out that the time of disintegration is not known *a priori* and that an observation is necessary in order to know it—an observation that changes the state of our knowledge of the system. Einstein admits that this answer may perhaps appear adequate when one is concerned only with a system on the microscopic level, such as a radioactive nucleus; but he adds that this is no longer so if one examines a case studied by Schrödinger.

One can, in fact, consider a system comprising not only an α radioactive nucleus, but also a measuring apparatus of macroscopic proportions, such as a Geiger counter, with an automatic recording mechanism. This apparatus may involve a paper strip with a clockwork drive, on which a mark is left when the counter operates. One then has a very complex system whose configuration space involves a very great number of dimensions, but there is nothing logically to prevent us from considering it. If we consider all the possible configurations after a very long time in comparison with the period of the radioactive atom, there will be *at most* one mark on the recording-strip. But since the current theory gives only the probability of the configurations, we can calculate only the relative probabilities of the position of the mark on the recording-strip. Now, Einstein observes, the position of the mark on the strip is a fact that belongs to the domain of Macroscopic Physics—which is not true of the instant of disintegration. If, thus, we consider the present quantum theory as giving a complete description of an individual system, we are forced to admit that the position of the mark on the strip is not a thing that belongs to the system *as such*, but that this position depends essentially on the observation that is made of the strip. Einstein considers this conclusion as highly unlikely.

By studying the transition from Classical to Wave Mechanics, Einstein finds still another argument against the purely probabilistic interpretation. Let us, along with Einstein, consider the motion in a field-free region of a macroscopic body which, to be as precise as possible, we will assume to be a homogeneous sphere of mass M. In Wave Mechanics, just as in Classical Mechanics, it can be shown that the motion of the center of gravity of the system (here, the center of the sphere) is the same as that of a point-mass of mass M. It is thus represented by the propagation of a wave-train obeying the equation of the propagation of Ψ waves with a value M for the mass. At the beginning instant, $t = 0$, this wave-train will have a form that will represent our uncertainties concerning the initial value of the coordinates of the center of gravity. At the end of a very long time t, the wave-train will be spread out, since the uncertainties concerning the coordinates of the center of gravity have increased. If, at that instant, we take a stereoscopic photograph of the body in motion, we will be able to deduce from this photograph with very great precision (compatible, none the less, with the uncertainty relations) the value of the coordinates of the center of gravity. The Ψ wave-train thus finds itself suddenly reduced by quite an appreciable amount. Now, the present interpretation of this reduction in the packet of probability is that the reduction is a result of the action of the measuring process. But here that interpretation is unacceptable, for the beams of light that illuminated the body at the moment it was photographed most certainly could not have exerted any appreciable action on the body, whose mass M may be quite large. This is another very strong objection to the current interpretation.

6. Conclusion

I have made a point of citing a few of the objections to the current interpretation of Wave Mechanics. As one can see, they emanate from some of the greatest scientific minds of our time.[8] For that reason alone, it is not entirely pointless to re-examine the only attempt that has been made to avoid the purely probabilistic interpretation—namely, the attempt I made in 1927 under the title of the "Theory of the Double Solution". Certain recent works, moreover, have, as we will see, once more drawn attention to that attempt.

[8] We may also point out a recent article by Schrödinger entitled "Are there quantum jumps?" (*Brit. J. Phil. Soc.*, 3 (1952) 11) in which there are some very interesting observations, especially those pertaining to the necessity of always considering limited wave-trains and of analyzing what happens along their boundaries.

PART TWO

The Theory of the Double Solution

Chapter VIII

INTRODUCTION AND PROGRAM

1. History of the theory of the Double Solution

At the time when I conceived the first ideas of Wave Mechanics, I was convinced that it was imperative to accomplish a fusion of the physical notions of waves and particles. I was, of course, aware that this would necessitate the introduction of a certain number of ideas foreign to Classical Physics, such as Planck's quantum of action; but I did not think that, as a result, one had to give up the kind of explanation found in Classical Physics; and above all, I did not think one had to forego the desire of achieving a clear representation of the physical world in the framework of space and time. So I sought to represent the wave-particle dualism to myself by a picture in which the particle would be the center of an extended phenomenon. This idea is found again and again in my early works.

Between 1924 (the date of publication of my Doctoral Thesis) and 1927, my ideas on the subject took on progressively more definite shape in a series of Notes to *Les Comptes rendus de l'Académie des Sciences*. And then they were summarized under the title of "The Theory of the Double Solution" in an article in the *Journal de Physique* [1].[1] My point of departure was this: The plane monochromatic wave that in my early works I had associated with the uniform and rectilinear motion of a free particle, as well as the Ψ wave of the continuous type employed by Schrödinger, and the continuous waves of the wave theory of light, all seemed to me to represent correctly the "phase" of the real wave phenomenon centered in the particle, but not its amplitude, which, to my way of thinking, involves a singularity that forms the particle in the strict sense of the word.

So I boldly laid down a hypothesis—that of the Double Solution—according to which the linear equations of Wave Mechanics admitted two kinds of solution: the continuous Ψ solutions one normally thinks of—the statistical nature of which was beginning to become clearly

[1] The bracketed numerals refer to the listings in the Bibliography at the end of the volume.

apparent at that time, thanks to the work of Born, and "singularity" solutions that would have a concrete meaning and be the true physical representation of the wave-particle dualism. Particles would then be clearly localized in space, as in the classical picture, but they would be *incorporated* in an extended wave phenomenon. For this reason, the motion of a particle would not obey the laws of Classical Mechanics according to which the particle is subject only to the action of forces exerted on it in the course of its trajectory, without experiencing any effect from the existence of obstacles that may be situated at some distance outside the trajectory. In my conception, on the contrary, the motion of the singularity was to be dependent on all the obstacles that hindered the free propagation of the wave phenomenon surrounding it, and there would result from this a reaction of the wave phenomenon on the particle—a reaction expressed in my theory by the appearance of a "quantum potential" entirely different from the potential of ordinary forces. And in this way the appearance of interference and diffraction phenomena would be explained.

Unfortunately the development of this theory of the Double Solution presented great mathematical difficulties. For that reason, when I was requested to present a paper on Wave Mechanics at the Solvay Physical Congress held in Brussels in October 1927, I contented myself with a presentation of my ideas in an incomplete and diluted form which I called the "pilot-wave theory" [2]. Here is what formed the basis of that truncated version of my conceptions: In my paper on the Double Solution, I had shown that the motion of the particle was defined, at least for the Newtonian approximation, by a formula that I have since called the "guidance formula"

$$\mathbf{v} = -\frac{1}{m} \operatorname{grad} \varphi \tag{1}$$

where φ is, to within the factor $1/\hbar$, the wave-phase of the singularity. Since, as I saw it, the phase must be the same for the wave containing the singularity as for the continuous Ψ wave, one could thus say that the particle was "guided" by the Ψ wave according to formula (1). And I used the term "pilot-wave theory" for the theory limited to the postulation of the existence of the particle and the Ψ wave, with no further reference to a wave containing a singularity [2]. This watered-down version of my original conception happened to coincide exactly with the one put forward at the same date by Madelung in his hydrodynamical interpretation of Wave Mechanics, but this simplified version had far less interest and profundity than my initial ideas on the Double Solution. My presentation at the Solvay Congress was received unfavorably, and the purely probabilistic interpretation of Bohr, Born and Heisenberg supported by Pauli, Dirac and others, was very clearly the one preferred by most of the scientists present. Pauli, in particular, criticized my theory by citing the example of Fermi's quantized rigid rotator.[2]

The objections raised against my approach, as well as the almost unanimous acclaim of the members of the Congress for the Bohr-Heisenberg interpretation (except for Lorentz, Schrödinger and Einstein, who raised the objection set forth in Chapter VII, section 2) made a very great impression upon me. In addition, when, after the conclusion of the Congress, I thought over the pilot-wave theory I had maintained, I became aware that it could not really furnish a concrete picture, in conformity with the conceptions of the older Physics, of the wave-particle dualism. It assumes, as a matter of fact, that the particle is guided in its motion by the propagation of the Ψ wave considered in Wave Mechanics. But that assumption could only lead to a concrete and causal theory of the type I was seeking if it were possible to consider this Ψ wave as an objective physical reality. But the Ψ wave usually employed in Wave Mechanics cannot be a physical reality; its normalization is arbitrary; its propagation, in the general case, is supposed to take place in an obviously fictitious configuration space, and the success of its probabilistic interpretation shows clearly that it is merely a representation of probabilities dependent upon the state of our knowledge and suddenly modified by every new piece of information. So I saw clearly that the pilot-wave theory could not supply the interpretation I sought; it did not achieve the clearcut separation of the objective and subjective, which had been given up by Bohr and his disciples, but which it was necessary to maintain if I was to arrive at a concrete and causal interpretation of Wave Mechanics.

On the other hand, my original theory of the Double Solution, by distinguishing the Ψ wave, with its probabilistic and subjective character, from the singularity-wave (u wave), which was to be a description of objective reality, might possibly supply the more classical type of interpretation I was after. But I knew only too well that the

[2] See Chap. XIV.

theory of the double solution likewise involved numerous difficulties, especially when it came to the existence and form of singularity-waves and to their relation to the Ψ waves, or when one had to interpret in terms of singularity-waves interference experiments of the Young-slit type, etc.

Confronted with all these difficulties, I gave up these attempts, for their outcome struck me as far too problematical. From 1928 on I embraced Bohr's probabilistic interpretation as the basis of my personal research, my teaching and my books.

During the summer of 1951, there came to my attention, much to my surprise, a paper by David Bohm which appeared subsequently in *The Physical Review* [3]. In this paper Bohm went back to my theory of the pilot-wave, considering the Ψ wave as a physical reality. He made a certain number of interesting remarks on the subject, and in particular, he indicated the broad outline of a theory of measurement that seemed to answer the objections Pauli had made to my approach in 1927.[3] My first reaction on reading Bohm's work was to reiterate, in a communication to the *Comptes rendus de l'Académie des Sciences* [4], the objections, insurmountable in my opinion, that seemed to render impossible any attribution of physical reality to the Ψ wave, and consequently, to render impossible the adoption of the pilot-wave theory. Takabayasi, moreover, subsequently took up these objections in papers where he developed certain aspects of Bohm's theory in an interesting way.

Jean-Pierre Vigier, who was doing research on unified theories in General Relativity, then called my attention to the great similarity between the demonstration I had given in 1927 in establishing the guidance formula (1) in the framework of the theory of the Double Solution and a demonstration indicated by Georges Darmois as early as 1926[4], developed independently by Einstein and Grommer [5] in 1927, and since taken up repeatedly in various forms by Einstein himself and by Fock, Infeld[6], Hoffmann, etc. The latter demonstration proves that a singularity in the gravitational field must, automatically, on account of the non-linear equations satisfied by the metric coefficients $g_{\mu\nu}$ of

[3] See Chap. XV.

[4] G. Darmois, *Les Équations de la gravitation einsteinienne*, *Mém. Sci. math.*, Gauthier-Villars, Paris, 1927.

[5] *Sitz. Preuss. Akad. Wiss.*, vol. I, 1927.

[6] *Rev. Mod. Phys.*, I, 21 (1949) 408.

space-time, follow a geodesic of the metric defined by the external gravitational field which is superposed on the field of singularity. Vigier, moreover, sought to clarify the similarity he had pointed out to me by suggesting that the wave of the particles present (naturally, the objective singularity-wave u) might in a certain sense determine the structure of space-time.

The similarity pointed out to me by Vigier struck me as of great interest, and I was led to explain in another Note [6] how one might try to resuscitate, not the theory of the pilot-wave (which still appeared to me to be unacceptable), but the theory of the Double Solution, which is not susceptible to the same objections in principle. In the conclusion of that note, I said that it contained only a program, and that the execution of this program would run into very great difficulties. Leaving aside the question of its compatibility with General Relativity, I would like, in the second part of the present work, to explain how far along the realization of that program is at the present time.

2. Problems to be treated in subsequent chapters

We will first of all review the results of our 1927 paper—in particular, the demonstration of the guidance formula and the introduction of the quantum potential. We will develop the resulting Dynamics for the point-singularity—a dynamics that, by means of the quantum potential, is found to depend upon boundary conditions imposed upon the wave phenomenon by the presence of obstacles. Then we will study some of the consequences of the formulas obtained, as well as a recent objection raised by Einstein in this connection.

An important point is the justification of the guidance formula and of the statistical meaning of the Ψ wave in the case of interacting systems of particles—a case where the Ψ wave considered in usual Wave Mechanics is supposed to be propagated in configuration space, which is an obviously fictitious space. From the causal point of view adopted by the Double Solution it must be demonstrated that the guidance formula and the statistical interpretation of Ψ both result from interactions between the singular regions of u-type waves evolving in three-dimensional physical space. In my article in the *Journal de Physique* of May 1927, I had outlined a demonstration of this sort in which *configuration space was considered as being formed by the coordinates of the singularities.* I thereby succeeded in representing the motion of the interacting particles as being effected in physical space, without being obliged to have recourse to configuration space. This fictitious space and the propagation of the Ψ wave in that space would then become merely tools for calculation convenient in making statistical predictions. By persevering in this line of attack, one ought to arrive at a physical interpretation of the use of both symmetrical and antisymmetrical Ψ wave functions in the Wave Mechanics of particle systems. The physical significance of Pauli's exclusion principle would become much clearer if we succeeded in showing that, for fermions, the u wave can involve only one singularity, while in the case of bosons several might be involved. I will set forth certain considerations relative to this subject which I have been able to develop and which seem to constitute a certain advance in that direction [7].

Next we will examine the important question of the statistical meaning of the quantity $|\Psi|^2$. In my 1927 paper I had pointed out that, as a result of the identity, postulated in the Double Solution, of the respective phases of the Ψ and u waves, the quantity $a^2 = |\Psi|^2$ obeyed the continuity equation

$$\frac{\partial a^2}{\partial t} + \operatorname{div}(a^2 \mathbf{v}) = 0 \tag{2}$$

$\mathbf{v}$ being the velocity defined by the guidance formula (1). Proceeding from this relation, I had shown that it was quite natural to suppose that $a^2 = |\Psi|^2$ gave the probability of the presence of the particle at a point when we do not know which of the possible trajectories is actually described by the particle. In this way we come back to the meaning that is currently associated with $|\Psi|^2$. The hypothesis made in order to reach that conclusion looks somewhat like the hypothesis made in Classical Statistical Mechanics when equal probability attached to equal volume elements in phase space is assumed solely on the basis of Liouville's theorem. But a justification seems called for that will be analogous to the ergodic demonstrations that we try to establish in Classical Statistical Mechanics. Bohm, in a recent paper [8], has come up with a line of argument that seems to lead to this justification.

Next we examine the objection which Pauli raised against the guidance formula in 1927, and this will be followed by an examination of the answer that can be made to it by relying on the basic idea of the limitation of wave-trains. The analysis of measuring processes has, in the overall consideration of these concepts, very great importance. By keeping the idea introduced as a result of the development of Quantum Mechanics, namely, that every measuring process, in general, modifies completely the state that existed before the measurement was made, we find that we are once more brought back to the Heisenberg uncertainty relations while at the same time reconciling them with the causal point of view. Such an analysis likewise permits us to discern the arbitrary postulate that forms the basis of both the "general theory of transformations"—so called in the usual treatises—and of von Neumann's arguments demonstrating the alleged impossibility of interpreting the probability laws of Wave Mechanics by means of hidden variables. In treating these questions we will reinforce our arguments, —especially with material from the papers of Bohm and Takabayasi already referred to.

Naturally, we must be able to extend all the foregoing considerations to the case of an electron with a spin obeying the relativistic equations of Dirac's theory (and even more generally to the case of particles with a spin greater than $\frac{1}{2}$). After a first inadequate attempt made by me in that direction, Vigier, using Bohm's ideas as his starting point, came up with a solution of this problem more satisfactory than mine, and it is his solution that will be presented [9]. It does not seem to me that the extension of the ideas of the Double Solution to the Dirac electron presents any special difficulties.

A very important point to be elucidated is the question of the nature of u solutions having a singularity of the kind postulated by the theory of the Double Solution. A certain number of reasons, which will be explained, and in particular the probable relationship between the theory of the Double Solution and the General Theory of Relativity, have led me to modify quite considerably the ideas I originally expressed in 1927. At that time I considered the u wave as a solution of the *linear* equation of propagation of Wave Mechanics, which would involve a singularity in the mathematical sense of the word.

It seems to me at the present time absolutely certain that the idea of singularity must be replaced by that of a very small singular region —in general, mobile—where the u function would take on very large values and obey a *non-linear* equation. Only outside this very small singular region would the u function approximately obey the linear equation of propagation of current Wave Mechanics. This new way of defining the u wave is in conformity with Vigier's ideas. Vigier thinks that in this way it would be possible to reconcile the theory of the

Double Solution with the ideas of Einstein—who always sought to represent particles by singular regions in the field—and perhaps also with the non-linear electromagnetism of Born.

The examination of the way in which eigen-values corresponding to quantized states appears within this framework of ideas, has recently led me to state more precisely the form of the waves and its relations to that of the Ψ waves [10]. The u waves would involve—in keeping, moreover, with a suggestion of Vigier's—outside the singular region in which they may take on very high values, a "regular" external part that would be proportional to the normalized Ψ wave, but with a multiplicative factor having a perfectly determined physical value—all of which would be in keeping with the objective character attributed to the u wave. In this way the Ψ wave ordinarily considered in Wave Mechanics and arbitrarily normalizable would continue to have the subjective and statistical character that it unquestionably possesses, but the external part of the u wave, which has an objective significance and a perfectly determined value, would be proportional to the Ψ wave normalized to one. Thus, without attributing a character of physical reality to the Ψ wave, the principal objection to the pilot-wave theory would be removed, and Bohm's point of view, *given the proper interpretation*, would become acceptable. In addition, as I will show, this conception of the u wave, shaped like a very high and very narrow pulse function with side extensions represented by a function proportional to the Ψ wave, seems capable of furnishing an explanation of interference experiments like that of the Young slits—an explanation that, twenty-five years ago, seemed to me an insurmountable obstacle in my attempt at finding a causal theory. The ensemble of ideas just summarized seems, thus, to make for considerable progress in the concepts of the theory of the Double Solution.

But there nevertheless remain very delicate points to be cleared up. It seems inadmissable, for a free particle associated with a group of almost monochromatic waves, that the outer regular portion of the u wave can be represented by the Fourier integral that represents the Ψ wave in the case of such a wave group. It is thus necessary to admit that the proportionality between the regular part of the u wave and the Ψ wave considered in the same problem by ordinary Wave Mechanics has limitations, and it becomes necessary, at least in certain cases, to "weaken" the connection established between the u and Ψ waves. The examination of an objection presented by Francis Perrin concerning the emission of particles by a point source as represented by a divergent spherical wave confirms this necessity. We will develop a suggestion according to which, from the non-linear character of the equation of propagation of the u wave and the rapid variation of u on the boundaries of the wave-train, there take place along the edges of the wave-train non-linear phenomena that would be impossible to predict with the usual linear equation of propagation.

Finally, we will study in the theory of the Double Solution the very important and very difficult problems related to the interpretation of reflection at a semi-transparent barrier, to the interpretation of the reduction of the probability packet, of the meaning of stationary waves, and the conservation of energy and impulse, etc. The success or failure of the causal interpretation of Wave Mechanics set forth in this work will, in the final analysis, depend on the possibility of overcoming the difficulties arising in these problems. It seems certain that if a way *is* found for overcoming these difficulties, it will be found most of all as a result of the finiteness of the wave-trains, and perhaps also as a result of the existence of non-linear phenomena on the wave-fronts and in the singular regions.

In conclusion, in order to simplify the terminology, I introduce the following convention: *In every instance where it is not essential to distinguish between the theory of the Double Solution and its degenerate form, the pilot-wave theory, I will employ the shortened term "causal theory".*[7]

[7] Most of the earlier, as well as the more recent, texts mentioned in this chapter will be found in a short work written by the present author in collaboration with Vigier [11].

Chapter IX

PRINCIPLES OF THE THEORY OF THE DOUBLE SOLUTION

1. General ideas

First let us summarize the general ideas that guided me when I conceived the theory of the Double Solution in 1927.

1. The wave-particle synthesis is to be achieved by representing the particle as a kind of singularity embedded in an extended wave-phenomenon.

2. The probabilistic interpretation of the continuous Ψ wave is on the whole correct and must be retained.

3. The Ψ wave being written in the form $ae^{\frac{i}{\hbar}\varphi}$ (with a and φ real), the phase φ has a profound physical significance corresponding to the indications of an infinite number of tiny clocks whose motion would be bound to that of the particle (see Chap. I, section 1). This phase φ must also be the phase of the singularity-wave which, in the theory to be constructed, will have to represent the real structure of the particle and of the wave phenomenon in which the particle is incorporated. On the other hand, the amplitude a of the Ψ wave, which is continuous, has no objective significance; it has only the character of representing a probability.

4. Of all the probabilities considered by standard Wave Mechanics, the probability of presence $|\Psi|^2$ has a sort of priority over all the others, for it corresponds in reality to the probability of the particle *being* at a point in the state represented by the Ψ wave. The other probabilities, such as $|c(\mathbf{k})|^2$ for the value $\hbar\mathbf{k}$ of the momentum, have a less immediate meaning; they are valid only after the action of an apparatus permitting the measurement of the quantity concerned, when one does not as yet know the result of that measurement.

5. Since every real phenomenon may be represented in the framework of space and time, it cannot be admitted that the problem of N particles interacting can be treated only by considering a propagation of waves in the system's configuration space, which is an obviously fictitious space. One should thus be able to state the problem, and even, in principle, to solve it, by considering N mutually interacting singu-

larity-waves propagated in physical, three-dimensional space. But one will then have to be able to demonstrate that the statistical result of the interaction is given exactly by consideration of the Ψ wave of the system in configuration space—a wave which, being only the representation of a probability, can be represented only in a fictitious framework.

2. The principle of the Double Solution

Equipped with these general ideas, I ventured to assume the following principle, to which I gave the name of "the principle of the Double Solution".

To every continuous solution $\Psi = ae^{\frac{i}{\hbar}\varphi}$ *of the equation of propagation of Wave Mechanics there must correspond a singularity solution* $u = fe^{\frac{i}{\hbar}\varphi}$ *having the same phase* φ *as* Ψ*, but with an amplitude* f *involving a generally mobile singularity.*

As I conceived it, the u function was the true representation of the physical entity "particle", which would be an extended wave phenomenon centered around a point (or an almost point-like region), which would constitute the particle in the strict sense of the word. I then considered the particle in the strict sense of the word as being defined by a true mathematical singularity, that is, by a point where the function f would become infinite. Considerations that will be referred to later lead me today to believe that the particle must be represented, not by a true point singularity of u, but by a very small singular region in space where u would take on a very large value and would obey a non-linear equation, of which the linear equation of Wave Mechanics would be only an approximate form valid outside the singular region. The idea that the equation of propagation of u, unlike the classical equation of Ψ, is in principle non-linear now strikes me as absolutely essential.

However, that may be, my conception of the particle being a continuous part of a wave phenomenon of which it forms the center seemed to make comprehensible the circumstance that the particle is localized and yet has a motion that can be influenced by the presence of obstacles situated far from it—as we seem forced to assume in order to interpret the existence of interference and diffraction phenomena.

Then what may be the significance of the continuous Ψ wave usually considered in Wave Mechanics? It must be merely a fictitious wave-function of subjective character, capable only of giving us information of a statistical order about the various possible motions of the particle, depending on whether it follows the one or the other of the trajectories defined by the phase φ.

In order to determine whether these general ideas could lead to acceptable results, I first examined the simplest case: that of the rectilinear and uniform motion associated with a plane monochromatic wave—a case which at an earlier date had given rise to my first reflections on Wave Mechanics.

3. The case of uniform rectilinear motion

Let us consider a particle in the absence of any field. In 1927, the *relativistic* equation for the propagation of the Ψ wave was assumed to be

$$\Box\Psi + \frac{m_0^2 c^2}{\hbar^2}\Psi = 0 \qquad (m_0 = \text{rest mass}). \tag{1}$$

At the present time we know, as was pointed out, that this equation applies only to particles of zero spin and that, for particles of non-zero spin we must use other forms of wave equations (for example, the Dirac equations for electrons of spin $\frac{1}{2}\hbar$), but this question will be neglected for the moment.

It is easily verified that equation (1) may have as solutions plane monochromatic waves of the form $\Psi = ae^{\frac{i}{\hbar}\varphi}$ with a constant and φ equal to

$$\varphi = Wt - \mathbf{p}\cdot\mathbf{r} \quad \left(W = \frac{m_0 c^2}{\sqrt{1-\beta^2}}; \quad \mathbf{p} = \frac{m_0 \mathbf{v}}{\sqrt{1-\beta^2}}\right). \tag{2}$$

Does there exist a singularity wave of the same phase φ, satisfying equation (1) and having the form $u = fe^{\frac{i}{\hbar}\varphi}$? By substitution in (1), we see that we must have $\Box f = 0$. Since that equation is invariant for a Lorentz transformation, we may write $\Delta f = 0$ by placing ourselves in the system appropriate to the particle where f does not depend on time. In this proper system, since the origin of the coördinates coincides with the particle, we will find as a spherically symmetrical solution

$$f(x_0, y_0, z_0) = \frac{C}{r_0} \quad \left(r_0 = \sqrt{x_0^2 + y_0^2 + z_0^2}\right) \tag{3}$$

and consequently

$$u(x_0, y_0, z_0, t_0) = \frac{C}{r_0} e^{\frac{i}{\hbar} m_0 c^2 t}. \tag{4}$$

The singularity-wave function having thus been obtained in the proper system, a simple Lorentz transformation will give the expression for it in another Galilean system and, on thus passing into the system in which the particle moves along the z-axis with velocity v, we will have [1]

$$u(x, y, z, t) = \frac{C}{\sqrt{x^2 + y^2 + \frac{(z - vt)^2}{1 - \beta^2}}} e^{\frac{i}{\hbar}(Wt - pz)}. \tag{5}$$

In this simple case of the absence of any field we have, thus, easily been able to find a solution with a mobile singularity verifying the principle of the Double Solution. Let me point out from the very outset that, if one wishes to replace the idea of a singularity by that of a very small singular region where u would obey a non-linear equation differing from the equation satisfied by Ψ, one would be led to consider the equation $\Delta f = 0$ as satisfied throughout the proper system, except in a very small finite region surrounding the origin. The spherically symmetrical solution $f = C/r_0$ could then represent f only outside a very small sphere surrounding the origin. Solution (5) would then be valid everywhere except in the interior of a very small ellipsoid surrounding the point $x = y = 0$, $z = vt$ flattened by the effect of the Lorentz contraction.

The singularity-solution u that we have just obtained has, in the proper system, the form of a polar-type solution with spherical symmetry. Dipolar-type solutions could likewise be found by putting, for example, $f(x_0, y_0, z_0) = Cx_0/r_0^3$ and similarly for N-polar-type solutions. The choices that can thus be made correspond to various hypotheses concerning the "internal structure" of the particle and the symmetry of that structure—a consideration that might be very important in a general theory of particles.

In my 1927 paper, I also indicated that other solutions of a different type existed. If, in fact, we place ourselves in the proper system, and if we seek, not a solution of the type

[1] Note that we could have added to (5) a solution $ae^{\frac{i}{\hbar}(Wt-pz)}$ with a having any value.

$$u = f e^{\frac{i}{\hbar} m_0 c^2 t_0} = f e^{i\omega_0 t_0} \qquad \omega_0 = 2\pi\nu_0$$

but solutions of the form

$$u = f e^{i\omega'_0 t_0} \quad \text{with } \omega'_0 \neq \omega_0$$

we find for f in the proper system the equation

$$\Delta f = \frac{1}{c^2} (\omega_0^2 - \omega_0'^2) f \tag{6}$$

and we are led to consider spherically symmetrical solutions tending to zero at infinity,

$$\begin{aligned} f(r_0) &= \frac{C}{r_0} \cos\left[\frac{r_0}{c}\sqrt{\omega_0'^2 - \omega_0^2} + C'\right] \quad &\text{for } \omega'_0 > \omega_0 \\ f(r_0) &= \frac{C}{r_0} e^{-\frac{r_0}{c}\sqrt{\omega_0^2 - \omega_0'^2}} \quad &\text{for } \omega_0 > \omega'_0. \end{aligned} \tag{7}$$

These solutions correspond to states of the particle in which its rest mass would not be equal to the constant m_0 which figures in the equation of propagation, but would have the value $\hbar\omega'_0/c^2$. These states, which I had at that time named "constrained" states of the particle, are interesting. They present an illuminating analogy with the circumstances one encounters for photons enclosed in a wave-guide whose motions correspond to rest masses which vary according to the form of the wave-guide and the type of waves propagated—rest masses much greater than the normal rest mass of the photon, which is zero or undetectably small.[2] Questions of this sort will come up again in a later chapter (Chap. XVII, section 6).

The foregoing results might be important if the causal theory ever succeeded in describing the structure of the particles and in predicting the values of their masses.

4. The interpretation of the Ψ wave in the case of uniform rectilinear motion

In the case we have just studied where we have the absence of both field and obstacles to propagation, we have found a solution with a singularity (or a singular region) moving with the velocity v along the z-axis.

[2] LOUIS DE BROGLIE, *Problèmes de la propagation guidée des ondes electromagnétiques*, 2nd edition, Gauthier-Villars, Paris, 1951, pp. 34–36.

We may imagine a great number of such particles whose motions are in parallel directions with velocity **v** and forming a sort of current of uniform density ρ. Let us then consider the wave $\Psi = ae^{\frac{i}{\hbar}\varphi}$ which has the same phase $\varphi = Wt - pz$ as the u wave of the particles. Since a is a constant, we can put

$$\rho = Ka^2. \tag{8}$$

We see, then, that the Ψ wave will represent the particle-motion characterized by the values W and **p** of energy and momentum by its phase, and it will thus represent by the square of its amplitude the density of the particles in space.

If we are dealing with only a single particle (we know that we must always be able to consider this case), and if we are in ignorance of which trajectories parallel to the z-axis it describes, and of its time description along that axis, it seems wholly natural to suppose, by reason of the equivalence of the parallel trajectories and of the various points of the same trajectory, that the probability of the particle's being at a point in space is everywhere the same. If we have taken the trouble to normalize the continuous Ψ wave, we will then be able to adopt as the expression of the probability

$$\rho = a^2 = |\Psi|^2. \tag{9}$$

And that is how the role of the continuous Ψ wave appears in the theory of the Double Solution; whereas the u wave with its singular region would describe the real structure of the particle, the continuous Ψ wave would be only a fictitious representation of the ensemble of possible positions of that particle.

Naturally, the unlimited u wave is an abstraction. In reality it would be necessary to consider limited trains of u waves, possessing in their central region a phase factor quite clearly identifiable with the phase of a plane monochromatic wave. The limited train of u waves should correspond to a packet of Ψ waves formed by the superposition of plane monochromatic waves destroying each other by interference outside the limits of the packet and identifiable in its central region with a plane monochromatic wave of constant amplitude. We will have occasion to return to this question later.

5. A study of the case of constant fields. Equations (J) and (C)

I will now, as I did in my 1927 article, study the case of constant fields, which is the simplest case after that of zero field.

We will assume that, in the reference system in which we are situated the field is static and is derivable from a potential $F(x, y, z)$. We will adopt as the wave equation valid in this case

$$\Box\Psi - \frac{2i}{\hbar} F(x, y, z) \frac{\partial\Psi}{\partial t} + \frac{1}{\hbar^2}\left(m_0 c^2 - \frac{F^2}{c^2}\right)\Psi = 0 \tag{10}$$

which is derived from the relativistic wave equation for a single Ψ—an equation given by formula (66) in Chapter II, when $A = 0$ and $eV = F$.

Let us imagine that the particle starts out by moving in a region R_0 of space where the field is zero, and then moves into a region R where the stipulated field of force is operative. In R_0, we may represent the particle by the wave $u = f(x, y, z, t)e^{\frac{i}{\hbar}(Wt-pz)}$ studied in section 3. The f function has a mobile singularity (or singular region) and φ coincides with the Hamiltonian action of the particle.

In order to obtain the representation of the particle in the field of force, we must continue the initial solution into the region R. In order to do this, we will still write $u = fe^{\frac{i}{\hbar}\varphi}$ with f and φ real, and carry that form over into equation (10), assumed valid for u. And here we come upon a most important circumstance that will arise for all the forms of the equation of propagation. Since f and φ are real, the wave equation will supply us, by separation of the real and the imaginary terms, with *two* distinct equations. One is the generalization of Jacobi's equation and will be designated as (J), the other is analogous to the continuity equation and will be designated as (C). In the present case, they have the following forms outside of the singular region:

$$\begin{aligned} &(\mathrm{J}) \quad \frac{1}{c^2}\left(\frac{\partial\varphi}{\partial t} - F\right)^2 - \sum_{x,y,z}\left(\frac{\partial\varphi}{\partial x}\right)^2 = m_0^2 c^2 + \hbar^2 \frac{\Box f}{f} \\ &(\mathrm{C}) \quad \frac{1}{c^2}\left(\frac{\partial\varphi}{\partial t} - F\right)\frac{\partial f}{\partial t} - \sum_{x,y,z}\frac{\partial\varphi}{\partial x}\frac{\partial f}{\partial x} = -\tfrac{1}{2} f \Box\varphi. \end{aligned} \tag{11}$$

It is easy to interpret equation (J). Since the field is static, the region R is analogous to a refracting medium with static properties, and, on entering that region, the wave will remain monochromatic with the frequency $\nu = W/h$, which it possessed in R_0. This means that the energy remains constant. We will then have in R,

$$\varphi(x, y, z, t) = Wt - \varphi_1(x, y, z); \quad \frac{\partial\varphi}{\partial t} = W; \quad \Box\varphi = -\Delta\varphi \tag{12}$$

and we can write equation (J) in the form

$$(\mathrm{J})\ \frac{1}{c^2}(\mathrm{W}-\mathrm{F})^2-\sum_{x,y,z}\left(\frac{\partial\varphi_1}{\partial x}\right)^2=m_0^2c^2+\hbar^2\frac{\Box f}{f}. \tag{13}$$

It is quite evident that, had we introduced into the wave equation, not the singularity wave u, but the corresponding wave $\Psi = ae^{\frac{i}{\hbar}\varphi}$ of the same phase φ, we would have found

$$(\mathrm{J}')\ \frac{1}{c^2}(\mathrm{W}-\mathrm{F})^2-\sum_{x,y,z}\left(\frac{\partial\varphi_1}{\partial x}\right)^2=m_0^2c^2+\hbar^2\frac{\Box a}{a}=m_0^2c^2-\hbar^2\frac{\Delta a}{a} \tag{14}$$

since a does not depend upon t in this case.

The comparison of (14) with (13) then shows that the principle of the Double Solution, which postulates the identity of the phases φ in u and in Ψ, results in:

$$\frac{\Box f}{f}=\frac{\Box a}{a}=-\frac{\Delta a}{a}. \tag{15}$$

At the very least, this relation is valid outside the singular region, where u and Ψ obey the same linear equation (10).

If the last terms of (13) and (14) (which will be of use later on in defining the quantum potential) are negligible, the approximation of Geometrical Optics will be valid; and then equations (J) and (J') become identical to the Jacobi equation of Relativistic Mechanics, and we see that φ_1, coincides with the shortened Jacobi function for static fields. To see this, one has only to put in equation (65) of Chapter III:

$$\mathrm{S}=\mathrm{W}t-\mathrm{S}_1, \qquad \mathrm{A}=0 \qquad \text{and } e\mathrm{V}=\mathrm{F}.$$

6. The guidance formula

In the approximation of the older Mechanics (Geometrical Optics), the velocity of the particle when it passes through a point M is determined by the fact that the momentum is equal to $(-\operatorname{grad}\varphi)_{\mathrm{M}} = (+\operatorname{grad}\varphi_1)_{\mathrm{M}}$, since then $\varphi = \mathrm{S}$ and $\varphi_1 = \mathrm{S}_1$.

This results from Jacobi's theory. We seek to establish that, *even outside the approximation of Geometrical Optics*, this relation still holds. We will thus obtain a sort of extrapolation of Jacobi's theory beyond the limits of validity of Geometrical Optics.

We will begin by stating that, for us, the particle in the strict sense of the word is defined by a very small mobile singular region, and it is then natural to suppose that, when we approach the center of this singular region, the function f increases very rapidly, undoubtedly as the inverse of a power of the distance from the center. Consequently, the derivative $\partial f/\partial s$ along the route followed must increase even more rapidly than f.

Let us consider a very small sphere S surrounding the singular region, in the middle of which f increases very rapidly. We will even assume, if the wave equation satisfied by u inside the singular region is not linear,

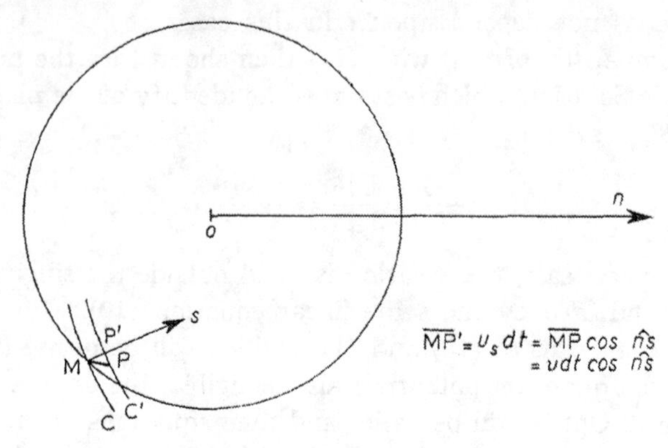

Fig. 6

that on S the u function still obeys the linear wave equation of standard Wave Mechanics. We will assume that on S the function φ_1 and its first derivatives have everywhere approximately the same value. This hypothesis amounts to supposing that the dimensions of the singular region are very small in comparison to a local wave-length corresponding to the phase φ_1. The amplitude f does not have, in general, the same value at all points of S, because that would be incompatible with the Lorentz contraction which the particle's structure undergoes by virtue of its motion.

Let us represent the sphere S with center O at the instant t.

Near a point M on this sphere let us imagine the positions at times t and $t + dt$ of the same surface $f =$ const. The direction **S** will be that of grad f in M, the direction **n** being that of grad φ_1, over the whole sphere. According to our hypotheses, we have

$$\left(f \Big/ \frac{\partial f}{\partial s}\right)_{M,t} \simeq 0. \tag{16}$$

Now, equation (C) allows us to write

$$\frac{1}{c^2}(\mathrm{W}-\mathrm{F})\frac{\partial f}{\partial t}+\frac{\partial f}{\partial s}\frac{\partial \varphi_1}{\partial n}\cos(ns) = -\tfrac{1}{2}f\Box\varphi. \tag{17}$$

Let us divide by $\partial f/\partial s$, taking (16) into account, and let us note that the displacement velocity of the value of f at M at the instant t is equal to $(-(\partial f/\partial t)/(\partial f/\partial s))_{\mathrm{M},t}$. There results;

$$v_s = \frac{c^2}{\mathrm{W}-\mathrm{F}}|\mathrm{grad}\,\varphi_1|\cos(ns). \tag{18}$$

Now, this result is valid at every point of the sphere S. Since $v_s = v\cos(ns)$, we conclude from this that the velocity $\mathbf{v}$ of the overall motion of the singular region takes place in the direction $\mathbf{n}$ of grad φ_1 and is given by the formula

$$\mathbf{v} = \frac{c^2}{\mathrm{W}-\mathrm{F}}\,\mathrm{grad}\,\varphi_1 = -\frac{c^2}{\mathrm{W}-\mathrm{F}}\,\mathrm{grad}\,\varphi. \tag{19}$$

This fundamental formula will be referred to as the "guidance formula". It shows that the overall motion of the singular region (that is, the particle's motion) is obtained quite simply by extrapolating the formula $\mathbf{p} = -\,\mathrm{grad}\,\mathrm{S}$ of the classical Jacobi's theory beyond the limits of Geometrical Optics, that is, of the older Mechanics. In fact, we have

$$\frac{\mathrm{W}-\mathrm{F}}{c^2} = \frac{m_0}{\sqrt{1-\beta^2}}$$

and formula (19) gives us

$$\mathbf{p} = \frac{m_0\mathbf{v}}{\sqrt{1-\beta^2}} = -\,\mathrm{grad}\,\varphi.$$

We know, moreover, that, when the Newtonian approximation is fulfilled ($v \ll c$), the dominant term of W is the rest energy m_0c^2 which turns out to be much greater than the potential energy F. We then have $\mathrm{W}-\mathrm{F} \simeq m_0c^2$, and the guidance formula takes the form

$$\mathbf{v} = -\frac{1}{m}\,\mathrm{grad}\,\varphi = \frac{1}{m}\,\mathrm{grad}\,\varphi_1 \tag{20}$$

m being the constant mass, equal to m_0, of Newtonian Mechanics.

Here we must make an important observation concerning the derivation of formula (19)—an observation that could lead to the prediction of phenomena not foreseen by the present interpretation. We have pointed out that our demonstration of the guidance formula assumes basically that the wave-length must be much greater than the dimensions of the singular region. If we designate by d the largest dimension of the singular region, the foregoing condition will be written

$$\frac{h}{p} \gg d. \tag{21}$$

Since in addition d seemingly must be of the order 10^{-13} cm, it is easily seen that for the ordinary particles of Atomic Physics, the condition may cease to be fulfilled only for velocities very close to c. One then has $p \simeq W/c$, and condition (21) will be written

$$W < \frac{hc}{d} = \frac{2 \times 10^{-16}}{d} \text{ ergs} \simeq 10^{-4} \text{ electron-volts (eV)}. \tag{22}$$

For $d \simeq 10^{-13}$ cm, this gives

$$W < 10^{9} \text{ eV}. \tag{23}$$

For particles of energy higher than 10^9 eV, the condition could not be fulfilled; in that case one would be confronted with an entirely novel situation, and the predictions of Wave Mechanics in its present form would no longer be verifiable.[3]

7. Introduction of the Ψ wave: its statistical significance

We will now introduce the Ψ wave for the case of static fields considered in the last two sections.

The phase φ_1 being assumed known, we can make an infinite number of possible motions correspond to it, which, according to (19), take place along various curves orthogonal to the surfaces $\varphi_1 = \text{const}$. Let us consider the totality of all possible motions of the particles along these trajectories. In this way we obtain the image of a cloud of particles. If we assume that at the beginning of their motion these particles were situated in a region R_0 where there was no field, the cloud will be represented statistically in R_0 by a plane and monochromatic Ψ wave (or, more precisely, by a limited wave-train approximated by a plane monochromatic wave along almost its entire extent) which will be associated with the totality of these rectilinear and uniform motions. Once they have reached the region R, the par-

[3] A different demonstration of the guidance formula will be found in the appendix.

ticles will have velocities given by the guidance formula, and their collective motion will be comparable to the stationary motion of the molecules of a fluid, since their velocity depends solely on their position and not on time, with the function φ_1 playing the part of a velocity potential.

The curves orthogonal to the curves $\varphi_1 =$ const. are here "lines of flow"; they form tubes in which the particles move forward. Since these tubes have a variable cross-section, the density ρ of the fluid varies from one point to the next, while remaining constant at every point in the course of time. From then on, the hydrodynamical equation of continuity, which the function $\rho(x, y, z, t)$ must satisfy, is reduced to

$$\operatorname{div}(\rho \mathbf{v}) = 0 \tag{24}$$

which, taking the guidance formula into account, may be written

$$\frac{\partial}{\partial n}\left(\log \frac{\rho}{\mathrm{W} - \mathrm{F}}\right) = -\frac{\Delta \varphi_1}{|\operatorname{grad} \varphi_1|}. \tag{25}$$

As in the case of the absence of any field, we will try to represent statistically the position of the particles in the cloud with the aid of a continuous Ψ wave. In the region R_0, the Ψ wave is, at the beginning of the motion, a limited wave-train approximating, along almost its entire extent, a plane monochromatic wave $\Psi = ae^{\frac{i}{\hbar}(\mathrm{W}t - \mathbf{p}\cdot\mathbf{r})}$ with a constant. As we have seen, it seems natural to define the density ρ as a function of the constant amplitude a by the relation $\rho = \mathrm{K}a^2$. Next, the Ψ wave-train will enter the region where the potential $\mathrm{F}(x, y, z)$ is operative; its propagation will then be governed by equation (10) and, according to the principle of the double solution, one should be able to express it in the form

$$\Psi(x, y, z, t) = a(x, y, z)\, e^{\frac{i}{\hbar}\varphi(x, y, z, t)}$$

φ being the same phase function as for u.

By substituting this form of Ψ in equation (10), one obtains by separation of the real from the imaginary terms, two relations (J′) and (C′). Equation (J′) has been written above [formula (14)] and leads us to relation (15) by comparison with (J). As for the relation (C′), since a does not depend on t, it takes on the simple form

$$-\operatorname{grad} a \cdot \operatorname{grad} \varphi + \tfrac{1}{2} a \,\Box \varphi = 0 \tag{26}$$

with

$$\varphi = \mathrm{W}t - \varphi_1 \quad \text{and} \quad \Box\varphi = \Delta\varphi_1$$

which can be written still another way

$$\frac{2}{a}\frac{\partial a}{\partial n} = \frac{\partial}{\partial n}(\log a^2) = -\frac{\Delta\varphi_1}{|\mathrm{grad}\ \varphi_1|}. \tag{27}$$

By comparing with (25), we find

$$\frac{\rho}{a^2(\mathrm{W} - \mathrm{F})} = \text{const.} \tag{28}$$

Since, in R_0, we have $\mathrm{F} = 0$ and $\rho = \mathrm{K}a^2$, we must have everywhere in R

$$\rho(x, y, z) = \mathrm{K}a^2(x, y, z)\left[1 - \frac{\mathrm{F}(x, y, z)}{\mathrm{W}}\right] \tag{29}$$

and the intensity a^2 of the Ψ wave determines ρ. In the Newtonian approximation where $\mathrm{F}/\mathrm{W} \simeq 0$, we find

$$\rho = \mathrm{K}a^2 = \mathrm{K}|\Psi|^2.$$

We may now return to the case of a particle by assuming that the cloud previously imagined represents an ensemble of possibilities. We assume, in fact, the initial velocity to be given in R_0 both in magnitude and direction; and, since we know nothing more, it seems permissible to suppose that the positions in the initial wave-train are equally probable. That being so, to every hypothesis on initial position there will correspond a well determined motion, and combining in our mind all these possibilities, we will obtain the equivalent of the motion of an infinitely dense cloud of identical particles. The probability of the particle's being present at a given instant in an element of volume $d\tau$ surrounding the point with coordinates x, y, z in region R is then given by $\rho(x, y, z)d\tau$ with the value (29) of ρ. In the Newtonian approximation, if we have taken care to normalize the Ψ function, the probability of presence of the particle will be

$$\rho(x, y\ z)d\tau = a^2(x, y, z)d\tau = |\Psi(x, y, z, t)|^2\, d\tau \tag{30}$$

and in this way we once more arrive at the usual statistical significance of $|\Psi|^2$.

The validity of the foregoing reasoning, with which I was satisfied in my 1927 paper, may be contested. We will return to this question in Chapter XIII.

8. The guidance formula and the theory of the pilot-wave

Let us point out immediately that, in the concept of the Double Solution, since the phase φ is assumed to be common to the u wave with the singular region and to the statistically continuous Ψ wave, the guidance formula (19) will permit us to deduce the trajectories of the particle from the sole knowledge of the Ψ wave. One may thus be tempted to speak no longer about the u wave and to consider the particle as a point mass whose existence would be postulated and whose motion, by hypothesis, would be determined (beginning with the phase of the continuous Ψ wave) by the guidance formula. It is this truncated, degenerate form of the theory of the Double Solution which I was unfortunately content to present at the Solvay Congress in October 1927 under the name of the "theory of the pilot-wave". In that form, my interpretation coincided with an attempt made around the same time by Madelung, often called the "Madelung hydrodynamical interpretation".

It does not seem that the theory, truncated in that way, is acceptable. If there exists—as the theory of the Double Solution assumes—an objective wave phenomenon represented by a wave u, having a singular region, whose propagation is modified by the action of external fields and by the presence of obstacles (interference and diffraction), one can then conceive that everything takes place as if the trajectory of the particle, which is really imposed upon it by the propagation of the u wave, were determined by the phase of the Ψ wave. But it is impossible to assume that it is the Ψ wave that regulates the motion of the particle because this Ψ wave is only a probability representation with a fictitious and subjective character.

We will be obliged to return repeatedly to this important question. We will see especially that, in his paper of January 1952, David Bohm has again taken up the theory of the pilot-wave, assuming that the Ψ wave is a "physical reality". This point of view seems inadmissible to me, even when we are concerned with the normalized Ψ wave of Wave Mechanics for a single particle, and then, all the more so when it is a matter of the Ψ wave of a system of particles in configuration space.

We will see later on, nevertheless (Chap. XVII), that, in the case of a particle, the external part of the u wave is probably (at least in a very great number of cases) proportional to the corresponding normalized Ψ wave with a completely determined proportionality constant. This permits us to establish a certain relation between the point of view of the Double Solution and that of Bohm. But of course, this relation does not hold for the case of the Ψ wave in a system in configuration space. We will go into all these matters much more thoroughly later on.

9. A study of the general case of non-static fields

In treating the general case of non-static fields, we again start from the relativistic equation of propagation of a single Ψ (valid for particles of zero spin). For a particle with an electric charge e moving in an electromagnetic field derivable from the scalar potential $V(x, y, z, t)$ and the vector potential $A(x, y, z, t)$, we have found [formula (66) of Chap. III] the following equation of propagation:

$$\Box\Psi - \frac{2i}{\hbar}\frac{e}{c^2}\frac{\partial\Psi}{\partial t} - \frac{2i}{\hbar}\sum_{x,y,z}\frac{e}{c}A_x\frac{\partial\Psi}{\partial x} + \frac{1}{\hbar^2}\left[m_0^2c^2 - \frac{e^2}{c^2}(V^2 - A^2)\right]\Psi = 0. \tag{31}$$

We will continue to assume that the particle under consideration begins by moving in a region R_0 of space where the field is absent and where the u wave has the form of a limited wave-train with a phase having almost everywhere the form $\varphi = Wt - \mathbf{p}\cdot\mathbf{r}$. Then the wave-train enters a region R in which the indicated electromagnetic field exists. We continue, in R, to look for a representation of the u solution of the form

$$u = f(x, y, z, t)e^{\frac{i}{\hbar}\varphi(x,y,z,t)}$$

where f and φ are real and where f involves a very small mobile singular region. By substituting this form in equation (31), we find the two equations (valid outside the singular region)

$$\begin{aligned}
&(\mathrm{J}) \quad \frac{1}{c^2}\left(\frac{\partial\varphi}{\partial t} - eV\right)^2 - \sum_{x,y,z}\left(\frac{\partial\varphi}{\partial x} + \frac{e}{c}A_x\right)^2 = m_0^2c^2 + \hbar^2\frac{\Box f}{f}\\
&(\mathrm{C}) \quad \frac{1}{c^2}\left(\frac{\partial\varphi}{\partial t} - eV\right)\frac{\partial f}{\partial t} - \sum_{x,y,z}\left(\frac{\partial\varphi}{\partial x} + \frac{e}{c}A_x\right)\frac{\partial f}{\partial x} = -\tfrac{1}{2}f\Box\varphi.
\end{aligned} \tag{32}$$

If the last term of (J) is negligible, Geometrical Optics is valid and φ coincides with the Jacobi function S of the Relativistic Dynamics of non-static fields. The departure from the older Mechanics is always bound up with the presence of the term $\Box f/f$.

In Classical Dynamics, one had to distinguish, in the presence of the vector-potential $\mathbf{A}$, the total momentum (Lagrangian momentum)

$$\mathbf{p} = -\operatorname{grad} S = \frac{m_0\mathbf{v}}{\sqrt{1-\beta^2}} + \frac{e}{c}\mathbf{A} \quad \left(\beta = \frac{v}{c}\right) \tag{33}$$

and the kinetic momentum[4]

$$\mathbf{g} = \frac{m_0 \mathbf{v}}{\sqrt{1-\beta^2}} = \frac{1}{c^2}\left(\frac{\partial S}{\partial t} - eV\right)\mathbf{v} \tag{34}$$

colinear with $\mathbf{v}$. One is thus led to put

$$\mathbf{g} = -\left[\text{grad } S + \frac{e}{c}\mathbf{A}\right]. \tag{35}$$

If we again take up the derivation of the guidance formula, starting from formula (32) (C), we find the overall velocity of the singular region to be

$$\mathbf{v} = -\frac{c^2\left(\text{grad } \varphi + \frac{e}{c}\mathbf{A}\right)}{\frac{\partial \varphi}{\partial t} - eV} = \frac{c^2 \mathbf{g}}{\frac{\partial \varphi}{\partial t} - eV}. \tag{36}$$

This formula corresponds exactly to the one we obtain, starting with (34)and (35), by extrapolating the Jacobi theory outside the domain of Geometrical Optics and by replacing S by φ.

Formula (19), valid for static fields, is obviously a special case of (36). In the Newtonian approximation, we may put $\partial\varphi/\partial t - eV \simeq m_0 c^2$ and we find

$$\mathbf{v} = -\frac{1}{m}\left(\text{grad } \varphi + \frac{e}{c}\mathbf{A}\right) = \frac{1}{m}\mathbf{g}. \tag{37}$$

Once more it is easy to introduce the continuous Ψ wave here, with its statistical meaning. It is still sufficient to consider first an infinite number of identical particles which would describe all the trajectories defined by the guidance formula on the basis of the function $\varphi(x, y, z, t)$ and which would form a sort of moving fluid.

The continuity equation would take on the form

$$\frac{\partial \rho}{\partial t} + \text{div } (\rho \mathbf{v}) = \frac{\partial \rho}{\partial t} + \text{div} \frac{\rho c^2 \mathbf{g}}{\frac{\partial \varphi}{\partial t} - eV} = 0. \tag{38}$$

Let us put $\rho' = \rho/(\partial\varphi/\partial t - eV)$, and let us take into account the relation between $\mathbf{g}$ and grad φ, as well as the Lorentz condition between

[4] Since $W = \partial S/\partial t = m_0 c^2/(\sqrt{1-\beta^2}) + eV$.

the potentials: $\partial V/\partial t + \operatorname{div} \mathbf{A} = 0$. There results

$$g\frac{\partial}{\partial n}\log \rho' + \frac{1}{c^2}\left(\frac{\partial \varphi}{\partial t} - eV\right)\frac{\partial}{\partial t}\log \rho' = -\Box\varphi \tag{39}$$

where $\mathbf{n}$ at every point defines the direction of motion.

As heretofore, we are going to try to associate the continuous Ψ wave with this stationary motion of the fluid. In the region R_0, this wave must be a limited wave-train approximating along practically its entire extent a plane monochromatic wave of phase $\varphi = Wt - \mathbf{p}\cdot\mathbf{r}$. In the region R, by virtue of the principle of the Double Solution, it will have to be written

$$\Psi(x, y, z, t) = a(x, y, z, t)e^{\frac{i}{\hbar}\varphi(x, y, z, t)}. \tag{40}$$

By introducing form (40) into equation (31), the separation of the real and imaginary terms supplies two relations, namely,

$$\begin{aligned} &(J') \quad \frac{1}{c^2}\left(\frac{\partial \varphi}{\partial t} - eV\right)^2 - \sum_{x,y,z}\left(\frac{\partial \varphi}{\partial x} + \frac{e}{c}A_x\right)^2 = m_0 c^2 + \hbar^2\frac{\Box a}{a} \\ &(C') \quad \frac{1}{c^2}\left(\frac{\partial \varphi}{\partial t} - eV\right)\frac{\partial a}{\partial t} - \sum_{x,y,z}\left(\frac{\partial \varphi}{\partial x} + \frac{e}{c}A_x\right)\frac{\partial a}{\partial x} = -\tfrac{1}{2}a\Box\varphi. \end{aligned} \tag{41}$$

The comparison of equation (J′) with equation (J) of formula (32) gives

$$\frac{\Box f}{f} = \frac{\Box a}{a} \tag{42}$$

a relation valid everywhere but in the singular region.

Equation (C′) may be written (n being understood to be in the direction of $\mathbf{v}$)

$$g\frac{\partial}{\partial n}\log a^2 + \frac{1}{c^2}\left(\frac{\partial \varphi}{\partial t} - eV\right)\frac{\partial}{\partial t}\log a^2 = -\Box\varphi. \tag{43}$$

By comparison with (39), we see that, if one follows the motion of the particles, the quantity ρ'/a^2 remains constant. Since in R_0 where $V = 0$, one has $\rho = Ka^2$, we deduce that in R,

$$\rho(x, y, z, t) = Ka^2\left(\frac{\partial \varphi}{\partial t} - eV\right). \tag{44}$$

In the Newtonian approximation where $\partial\varphi/\partial t - eV \simeq m_0 c^2$, one may write, if the Ψ function has been normalized,

$$\rho(x, y, z, t) = a^2 = |\Psi|^2. \tag{45}$$

Let us now suppose that there is only one particle, and let us assume, in the region R_0, that we may consider all the positions of the particle in the initial Ψ wave-train as equally probable. The cloud of particles considered above will then represent merely the ensemble of possible motions of the single particle corresponding to the phase function $\varphi(x, y, z, t)$ which we assume to be known. We conclude from this, as previously, that in the Newtonian approximation, the intensity $a^2 = |\Psi|^2$ of the normalized Ψ wave must give in absolute value the probability of finding the particle at the point x, y, z of the region R at the instant t. With the reservation that there may be brought against this sort of demonstration criticisms of the kind already pointed out, we thus, in the way just described, arrive at the significance usually attributed to $|\Psi|^2$.

Chapter X

THE DYNAMICS OF THE PARTICLE IN THE CAUSAL THEORY

1. The Lagrange and the Hamilton equations

If we consider equation (J) obtained in the preceding chapter [formula (32)], we see that it can take on the usual form of the relativistic Jacobi's equation by writing it

$$\frac{1}{c^2}\left(\frac{\partial\varphi}{\partial t} - eV\right)^2 - \sum_{x,y,z}\left(\frac{\partial\varphi}{\partial x} + \frac{e}{c}A_x\right)^2 = M_0^2c^2 \tag{1}$$

on condition that we attribute a *variable rest-mass* to the particle,

$$M_0 = \sqrt{m_0^2 + \frac{\hbar^2}{c^2}\left(\frac{\Box f}{f}\right)} = \sqrt{m_0^2 + \frac{\hbar^2}{c^2}\left(\frac{\Box a}{a}\right)} \tag{2}$$

which is a function of the particle's position and of time. It is expressed in terms of two equal quantities $\Box f/f$ or $\Box a/a$ which must be calculated in the immediate vicinity of the particle (on the sphere S, previously introduced).

The older Mechanics neglected the second term under the radical, which is tantamount to assuming that h is infinitely small.

Having established this, we will be able to derive the particle's motion from Hamilton's principle and to arrive once more at the Lagrange equations of Relativistic Dynamics, on condition that we include in them the variable rest-mass M_0. We will then define the Lagrange function by putting

$$\mathscr{L} = -M_0c^2\sqrt{1-\beta^2} - eV + \frac{e}{c}\mathbf{v}\cdot\mathbf{A} \quad \left(\beta = \frac{v}{c}\right) \tag{3}$$

and Hamilton's principle of stationary action will be written

$$\delta\int_0^t dt\,\mathscr{L} = 0. \tag{4}$$

We are then led, as usual, to the Lagrange equation,

$$\frac{d}{dt}\left(\frac{\partial\mathscr{L}}{\partial v_x}\right) = \frac{\partial\mathscr{L}}{\partial x}, \ldots \tag{5}$$

and the Lagrange momenta are introduced by the usual definitions.

$$p_x = \frac{\partial \mathscr{L}}{\partial v_x} = \frac{M_0 v_x}{\sqrt{1-\beta^2}} + \frac{e}{c} A_x, \ldots \tag{6}$$

The name "energy" is always used to designate the following quantity

$$W = \sum_{x,y,z} v_x p_x - \mathscr{L} = \frac{M_0 c^2}{\sqrt{1-\beta^2}} + eV \tag{7}$$

and it is easily verifiable that it remains constant in a static field. It is also verifiable that by putting

$$\frac{\partial \varphi}{\partial t} = W, \qquad -\frac{\partial \varphi}{\partial x} = p_x, \qquad -\frac{\partial \varphi}{\partial y} = p_y, \qquad -\frac{\partial \varphi}{\partial z} = p_z, \tag{8}$$

we arrive, for φ, at the generalized Jacobi's equation,

$$\frac{1}{c^2}\left(\frac{\partial \varphi}{\partial t} - eV\right)^2 - \sum_{x,y,z}\left(\frac{\partial \varphi}{\partial x} + \frac{e}{c} A_x\right)^2 = M_0 c^2. \tag{9}$$

In addition, by combining expressions (6), (7) and (8), we have no difficulty in arriving once again at the guidance formula.

It is thus seen that, when the u function (or the Ψ function) is known, the Hamilton-Lagrange theory makes it possible, in principle, to calculate the form of the possible trajectories as well as the motion of the particle.

One may try to see whether the equations of the Dynamics can be put in canonical Hamiltonian form. The Hamiltonian expression for the energy is here

$$H(x, y, z, p_x, p_y, p_z, t) = c\sqrt{M_0 c^2 + \sum_{x,y,z}\left(p_x - \frac{e}{c} A_x\right)^2} + eV. \tag{10}$$

As in the case for the constant rest-mass, one may verify the validity of the equations

$$\frac{dq_i}{dt} = \frac{\partial H}{\partial p_i} \qquad (i = 1, 2, 3). \tag{11}$$

We then obtain

$$\frac{dp_x}{dt} = -\frac{\partial \mathscr{L}}{\partial x} = -c^2\sqrt{1-\beta^2}\,\frac{\partial M_0}{\partial x} - e\frac{\partial V}{\partial x} + \frac{e}{c}\mathbf{v}\cdot\frac{\partial \mathbf{A}}{\partial x} \tag{12}$$

and

$$-\frac{\partial H}{\partial x} = -e\frac{\partial V}{\partial x} - \frac{c^2}{W - eV}\left[c^2 M_0 \frac{\partial M_0}{\partial x} + \sum_{x,y,z}\left(p_x - \frac{e}{c}A_x\right)\left(-\frac{e}{c}\frac{\partial A_x}{\partial x}\right)\right]$$

$$= -e\frac{\partial V}{\partial x} + \frac{e}{c}\mathbf{v}\cdot\frac{\partial \mathbf{A}}{\partial x} - c^2\sqrt{1-\beta^2}\frac{\partial M_0}{\partial x} = \frac{dp_x}{dt}. \tag{13}$$

The second group of Hamilton's equations

$$\frac{dp_i}{dt} = -\frac{\partial H}{\partial q_i} \qquad (i = 1, 2, 3) \tag{14}$$

is thus also valid.

2. The relativistic formalism of the foregoing Dynamics

It is easy to transcribe the foregoing by utilization of the formalism of General Relativity. The wave equation (1) is then written in the form

$$\frac{1}{\sqrt{-g}}\frac{\partial}{\partial x^k}\left[\sqrt{-g}\, g^{kl}\frac{\partial \Psi}{\partial x^l}\right] - \frac{2i}{\hbar}P^k\frac{\partial \Psi}{\partial x^k} + \frac{1}{\hbar^2}\left(m_0^2 c^2 - \frac{e^2}{c^2}P^2\right)\Psi = 0 \tag{15}$$

where, since the g_{ik} are the classical coefficients of the metric of space-time, the g^{ik} are the corresponding contravariant components and g the determinant of the g_{ik}. The P^k are the components of the "four-vector potential" equal to A_x, A_y, A_z and V and we have

$$P^2 = P \cdot P = P_k P^k = V^2 - A^2.$$

As for the operator

$$\frac{1}{\sqrt{-g}}\frac{\partial}{\partial x^k}\left[\sqrt{-g}\, g^{kl}\frac{\partial}{\partial x^l}\right],$$

it is the generalization of the Dalembertian:

$$\square = \frac{1}{c^2}\frac{\partial^2}{\partial t^2} - \Delta.$$

If we put

$$\Psi = a(x_1, x_2, x_3, x_4)e^{\frac{i}{\hbar}\varphi(x_1, x_2, x_3, x_4)} \tag{16}$$

with a and φ real, by adopting definition (2) of M_0, we find by substitution of (16) in (15)

$$\text{(J)} \quad g^{kl}\left(\frac{\partial \varphi}{\partial x^k} - eP_k\right)\left(\frac{\partial \varphi}{\partial x^l} - eP_l\right) = M_0^2 c^2$$

$$\text{(C)} \quad \frac{1}{\sqrt{-g}}\frac{\partial}{\partial x^k}\left[\sqrt{-g}\, g^{kl} a^2\left(\frac{\partial \varphi}{\partial x^l} - eP_l\right)\right] = 0. \tag{17}$$

The four-vector velocity with the components $u^l = dx^l/ds$ will be given by the formula

$$M_0 c u^l = g^{kl}\left(\frac{\partial\varphi}{\partial x^k} - eP_k\right) \tag{18}$$

which is here the expression of the guidance formula. Clearly one has $u^l u_l = 1$.

Equation (C) tells us that the four-vector whose covariant components are $a^2(\partial\varphi/\partial x^k - eP_k)$ has a zero divergence. We can thus assume that this vector is proportional to the "current density" four-vector $C^l = \rho_0 u^l$ of the cloud particles previously considered, and we will write

$$C^l = \rho_0 u^l = K g^{kl} a^2 \left(\frac{\partial\varphi}{\partial x^k} - eP_k\right) \tag{19}$$

which, by reason of (18), will become

$$\rho_0 = KM_0 c a^2. \tag{20}$$

The number of particles in the cloud per unit-volume is given by the fourth component of the four-vector C^l, that is, according to (17), (19) and (20),

$$\rho = C^4 = KM_0 c a^2 u^4 = K a^2 g^{4k}\left(\frac{\partial\varphi}{\partial x^k} - eP_k\right). \tag{21}$$

If there is no gravitational field, one can give the g_{ik} their well-known Galilean values ($g_{11}^{(0)} = g_{22}^{(0)} = g_{33}^{(0)} = -1$, $g_{44}^{(0)} = 1$ and $g_{ik}^{(0)} = 0$ for $i \neq k$). We then once more arrive at the formulas previously obtained.

We may also note that *in the absence of electromagnetic and gravitational fields* the principle of least action will be written

$$\delta \int dx^i M_0 c u_i = c\delta \int ds M_0 = 0 \tag{22}$$

where the integral is taken along the world line and the variation at the end points of the world line is zero.

Space-time being Euclidean in that case, we will put $d\sigma = (M_0/m_0)ds$, whence

$$d\sigma^2 = \frac{M_0^2}{m_0^2} ds^2 = \frac{M_0^2}{m_0^2} g_{ik}^{(0)} dx^i dx^k = \gamma_{ik} dx^i dx^k \tag{23}$$

with the definition

$$\gamma_{ik} = \frac{M_0^2}{m_0^2} g_{ik}^{(0)} = \left(1 + \frac{\hbar^2}{m_0^2 c^2} \frac{\Box a}{a}\right) g_{ik}^{(0)} = \left(1 + \frac{\hbar^2}{m_0^2 c^2} \frac{\Box f}{f}\right) g_{ik}^{(0)} \quad (24)$$

where the $g_{ik}^{(0)}$ are the Galilean values of g_{ik} pointed out above.

Thus, even if the particle is not subjected to any gravitational or electromagnetic field, its possible trajectories, as anticipated in the theory of the Double Solution, and defined by $\delta \int ds = 0$, are the same as if space-time possessed non-Euclidean metrics defined by γ_{ik}.

In other words, if interference or diffraction phenomena arise, brought about by limiting conditions, and give a non-zero value to the quantity $\Box a/a = \Box f/f$, everything takes place as if space-time possessed, as far as the particle is concerned, a non-Euclidean metric defined by γ_{ik}. This observation is related to Vigier's current attempts to link the theory of the Double Solution to the conceptions of General Relativity by assuming that the metrics of space-time really depend at every point on the local value of the u function.

3. The "Quantum potential" and its interpretation

If, with the help of a calculation that is entirely classical in Relativity Theory, we determine the geodesics corresponding to the line element $d\sigma^2 = \gamma_{ik} dx^i dx^k$, we find that these geodesics have the equations

$$\frac{d}{ds}(M_0 c u_l) = c \frac{\partial M_0}{\partial x^l} \quad (l = 1, 2, 3, 4). \quad (25)$$

The equations give us once more exactly the same motion of the particle as obtained above.

Form (25) of the equations of motion is most instructive, because it shows us that, if M_0 were a constant, the components $M_0 c u_l$ of the momentum would be constant in the absence of any field. If these components generally vary, that is because the variations of M_0 are equivalent to the existence of a "Quantum field" represented by the second member of equations (25). This quantum field may exist, even in the absence of any field of the classical type (gravitational or electromagnetic); it arises the moment that limiting conditions bring about the appearance of interference and diffraction phenomena and give a non-zero value to the quantity $\Box f/f = \Box a/a$ at the point where the particle is located. The quantum field expresses, in a way, the reaction of the u wave, distorted by the obstacles that hinder its free propagation, on the particle that is embedded in that wave in the form of a small singular region.

To make things as clear as possible, let us assume the validity of the Newtonian approximation. In it we neglect the terms in β^2 and, in addition, consider the second term under the radical in the expression of M_0 to be very small in comparison with the first, and this permits us to write

$$M_0 \simeq m_0 + \frac{\hbar^2}{2m_0 c^2}\left(\frac{\Box f}{f}\right) = m_0 + \frac{\hbar^2}{2m_0 c^2}\left(\frac{\Box a}{a}\right) \tag{26}$$

where the parentheses are taken to mean that the quantities are evaluated on the boundaries of the singular region on the sphere S.

Thenceforward we will have for the components of momentum and energy, in the absence of any field of the classical type, the expressions

$$p = m_0 v \qquad W = m_0 c^2 + \tfrac{1}{2} m_0 v^2 + Q \tag{27}$$

with

$$Q = \frac{\hbar^2}{2m_0}\left(\frac{\Box f}{f}\right) = \frac{\hbar^2}{2m_0}\left(\frac{\Box a}{a}\right). \tag{28}$$

Since in the Newtonian approximation one can neglect the term $(1/c^2)(\partial^2 a/\partial t^2)$ in $\Box a$, we will also have the expression pointed out by Bohm:

$$Q = -\frac{\hbar^2}{2m}\frac{\Delta a}{a} = \frac{\hbar^2}{4m}\left[\frac{\Delta a^2}{a^2} - \frac{(\text{grad}\ a^2)^2}{2a^4}\right] \tag{29}$$

where only a^2 is involved.

Q is the "quantum potential" whose quantum field is the negative gradient of this potential. The Lagrange equations then take on the simple form

$$\frac{dp_x}{dt} = -\frac{\partial Q}{\partial x} \quad \ldots \quad \frac{d\mathbf{p}}{dt} = -\,\text{grad}\ Q. \tag{30}$$

Naturally, if an external field existed which acted upon the particle, it would be necessary to add, in expression (27) for W, a potential-energy term of the classical type from which a field, in the older sense, would be derived. And this field would be added to the quantum field, —grad Q, in the second member of equations (30).

The physical origin of the quantum potential and of the field derived from it is now quite clear. When there are no external fields of the classical type, the quantity $\Box f/f = \Box a/a$ may become non-zero as a result of the presence of obstacles (in mathematical language due to boundary conditions) which modify the propagation of the *u* wave and cause phenomena of the interference or diffraction type to appear. In this way the constraint imposed upon the wave phenomenon, which surrounds the particle embedded in the wave as a small singular region, then reacts on the motion of this singular region and gives rise to the appearance of complicated trajectories determined by the guidance formula—trajectories from which interference and diffraction phenomena as we know them would result. Such is the point of view—certainly a most attractive one—of the theory of the Double Solution.

As I pointed out some twenty-five years ago, one is in this way brought back to a very old intuitive interpretation of diffraction phenomena—the interpretation, in fact, that the partisans of the old corpuscular theory of light, from Newton through Biot and Laplace, maintained. "If," they said, "light is deflected when it passes close to the edge of a screen, it must be because the edge of the screen exerts a force on the particle that makes it deviate from its normal rectilinear path." With the notion of the quantum potential we can similarly say: "If light is diffracted by the edge of a screen, it must be because the *u* wave of the photon is hindered in its propagation by the edge of the screen and a reaction on the photon's motion results; this reaction is explained by the intervention of the quantum potential and has as its effect the curvature of the photon's trajectory." This interpretation, so attractive because of its concrete character, would naturally also be valid for the diffraction of a particle other than a photon—for example, the diffraction of an electron by the edge of a screen (the Börsch phenomenon).

It is well to insist on the fact that the quantum potential appears every time we leave the domain of Geometrical Optics; that is, not only when obstacles placed in the path of the wave produce phenomena of the interference and diffraction type, but also when very rapid variations of classical fields in space no longer allow the validity of Geometrical Optics for the propagation of the wave (the case of the electron in an atom).

One must also note that the variable rest mass M_0 and the quantum potential deduced from it are not functions of x, y, z, t above, but also functionals of boundary conditions.

As we have just seen, the quantum potential is contained in the variation of the rest mass M_0, and, when the approximation is valid, it is expressed by the quantity Q of formula (28) which we may express as we choose by either of the two equal expressions $\Box f/f$ or $\Box a/a$. One may thus be tempted to define the quantum potential solely by proceeding from a, that is, from the continuous wave. In that way one is brought back to the theory of the pilot-wave which I put forth at the Solvay Congress of 1927 and which was taken up again by Bohm in his 1952 article.

But, as we know, that point of view hardly seems acceptable. If we wish to return to a causal theory, it would be necessary, in order to define the quantum potential by $\Box a/a$, to attribute a physical reality to the Ψ wave in the way Bohm sought to; and we have seen that this seems impossible. The reaction of the wave upon the particle, analytically represented by the quantum potential, cannot in actuality arise from the Ψ wave, which is fictitious. It must arise from the u wave representing the physical phenomenon of a wave type, in which, in the theory of the Double Solution, the particle is embedded in the form of a singular region.

The true physical expression of the quantum potential must thus be the one that introduces $\Box f/f$, and not the one introducing $\Box a/a$. It is only because the conditions imposed by the hypothesis of the Double Solution lead us to assume the equality of these two relations that we can entertain the illusion that the quantum potential arises from the Ψ wave.

Nevertheless, one may find rather peculiar this coincidence of the formulas of the solution and those of the pilot-wave—a coincidence already observable in the guidance formula. A great deal of light will be thrown on this matter in the study to be made further on (Chap. XVII) of the form of the u wave outside the singular region.

Chapter XI

A FEW CONSEQUENCES OF THE GUIDANCE FORMULA

1. The stationary states of the hydrogen atom

We wish to study a few of the consequences of the guidance formula, and we begin with the study of the stationary states of the hydrogen atom, limiting ourselves to the Newtonian approximation.

The stationary states of the H atom are characterized by three quantum numbers n, l, m. The principal quantum number n defines quantized energy by the classical Bohr formula, $E_n = -2\pi m_0 e^4/n^2 h^2$, where m_0 is the mass of the electron; the quantum number l defines the square of the angular momentum **M** by the formula $\mathbf{M}^2 = l(l+1)\hbar^2$; and finally, the quantum number m is associated with the z-component of **M** by the formula $M_z = m\hbar$. One thus sees that each quantized state (n, l, m) is defined in relation to a certain z-axis. Between the three quantum numbers we have the relations $0 \leqq l \leqq n-1$ and $-l \leqq m \leqq l$.

The wave function corresponding to the stationary state (n, l, m) is

$$\Psi_{nlm}(r, \vartheta, \varphi, t) = F_{nl}(r, \vartheta) e^{im\varphi} e^{\frac{i}{\hbar} E_n t} \tag{1}$$

with F_{nl} real. The phase, which I here designate by Φ in order to avoid confusion with the longitudinal angle φ, is $E_n t - \hbar m\varphi = \Phi$. The guidance formula shows that in state (1) the velocity of the electron must be in a plane perpendicular to the z-axis and that it is equal to

$$v_\varphi = -\frac{1}{m_0}\frac{1}{r}\frac{\partial \Phi}{\partial \varphi} = -\frac{1}{m_0 r} m\hbar \qquad v_r = v_\vartheta = 0 \tag{2}$$

whence

$$M = M_z = m_0 r v_\varphi = m\hbar \quad (-l \leqq m \leqq l). \tag{3}$$

The angular momentum of the electron around the z-axis is thus an integral multiple of $\hbar$—a classical result. In state (1), the electron would be accelerated in a uniform circular motion around the z-axis.[1]

In the special case of the ground state, we have $n = 1$, $l = m = 0$, from which $M^2 = M_z = 0$. The H atom thus has no angular momentum in its fundamental state, which is another classical result of Wave Mechanics. We find then

$$\mathbf{v} = 0. \tag{4}$$

The electron would be motionless at a point inside the atom. This latter result would be equally valid for all the s states when $l = m = 0$ [and even for all the other states when we assume that the Ψ given by (1) is replaced by one of the linear combinations which cause the real functions $\sin m\varphi$ or $\cos n\varphi$ to appear in place of the imaginary exponential $e^{im\varphi}$].

How can the electron remain motionless when it is subjected to the Coulomb force of the nucleus? The answer is very easy to give if we remember that, in the causal theory, the electron is also subjected to the quantum force derived from the quantum potential, for this quantum force exactly counterbalances the Coulomb force. Let us verify this for the H atom whose Ψ wave function has the form

$$\Psi_{100} = Ce^{-\frac{r}{r_0}} e^{\frac{i}{\hbar} E_0 t} \tag{5}$$

where r_0 is the radius of the circle K in the original Bohr theory and is equal to $\hbar^2/me^2$. Here the amplitude a of Ψ is then

$$a(r) = |C| e^{-\frac{r}{r_0}} \tag{6}$$

and we find

$$\frac{\Delta a}{a} = \frac{1}{a}\left(\frac{\partial^2 a}{\partial r^2} + \frac{2}{r}\frac{\partial a}{\partial r}\right) = \frac{1}{r_0}\left(\frac{1}{r_0} - \frac{2}{r}\right) \tag{7}$$

from which we obtain, for the quantum potential Q,

[1] Since the Ψ wave has form (1), all the circles centered at the origin in the equatorial plane are possible trajectories according to the guidance formula. Now, among these circles are found the circular trajectories with quantized radii of Bohr's original theory. On these Bohr circles, the equation $m_0(v^2/r) = e^2/r^2$ must be valid; and, in the causal theory, the same equation, with inclusion of the term $-dQ/dr$ in the second member, where Q is the quantum potential, must likewise be valid. Whence the following curious result: The circular trajectories of Bohr are characterized in the equatorial plane by the fact that the quantum potential possesses for this value of r a maximum or a minimum.

$$Q = -\frac{\hbar^2}{2m_0}\frac{\Delta a}{a} = -\frac{\hbar^2}{2m_0}\frac{1}{r_0}\left(\frac{1}{r_0} - \frac{2}{r}\right) \tag{8}$$

m_0 always being the electron's mass.

The quantum force is radial and has as a value, because of the value of r_0,

$$F_Q = -\frac{\partial Q}{\partial r} = -\frac{\hbar^2}{m_0 r_0}\frac{1}{r^2} = \frac{e^2}{r^2}. \tag{9}$$

Since the Coulomb potential is $V = -e^2/r$, the Coulomb force is radial and equal to

$$F_C = -\frac{\partial V}{\partial r} = -\frac{e^2}{r^2} = -F_Q; \tag{10}$$

F_C and F_Q are thus indeed equal and opposite and constitute an equilibrium.

It is easy to obtain the value of the quantized energy for the fundamental state, for we have

$$\begin{aligned} E &= \text{kinetic energy} + V + Q \\ &= 0 - \frac{e^2}{r} - \frac{\hbar^2}{2m_0}\frac{1}{r_0}\left(\frac{1}{r_0} - \frac{2}{r}\right) = -\frac{\hbar^2}{2m_0 r_0^2} = -\frac{1}{2}\frac{m_0 e^4}{\hbar^2}. \end{aligned} \tag{11}$$

Moreover, one can immediately obtain this same value by proceeding from Schrödinger's equation,

$$\Delta\Psi + \frac{2m_0}{\hbar^2}\left(E + \frac{e^2}{r}\right)\Psi = 0. \tag{12}$$

Since in every s state we have $\Psi = a(r)e^{\frac{i}{\hbar}Et}$ with a real, we immediately read from equation (12)

$$E = V + Q \tag{13}$$

which brings us back to the result (11).

Let us make a few comments on the results obtained, which may appear strange from various points of view.

First of all, if in the s states, the electron is motionless inside the atom, its position together with that of the nucleus defines a special axis—a fact that may appear contrary to the isotropy of the ground state. Now, the isotropy envisaged by ordinary Wave Mechanics (for $|\Psi|^2 = e^{-\frac{2r}{r_0}}$ has spherical symmetry) is definitely verified by experiment. In addition, it is difficult to conceive how, in the s states where the electron would be motionless, or even in the more general case of a state given by (1) where the electron would have a uniform circular motion, the statistical distribution $|\Psi|^2 = |F_{nl}(r, \vartheta)|^2$ of the probabilities of presence could be realized.

One might be tempted to answer these objections in the following way: First, if in the *s* states the electron is motionless at a point M, the straight line OM joining the nucleus to that position may be oriented in space in any way whatsoever, so that for a collection of atoms in the *s* state, there will indeed be a statistically symmetrical spherical distribution around the nucleus, and it is that distribution which will in general be revealed by experiment. Nor must one forget that in the causal theory, just as in the usual interpretation, the angular momenta of the *s* states is zero, contrary to the original Bohr theory, so that the experiments which seem to demonstrate the zero value of the angular momentum of *s* states are not in contradiction with the causal theory.

As for the realization of the probability of presence $|\Psi|^2$, it is essential to note that a stationary state of the form (1) must be considered exceptional; in general, the Ψ function will be a superposition of such states, so that one will have $\Psi = \sum_{nlm} c_{nlm}\Psi_{nlm}$. Then the phase Φ will take on a very complicated form, and the guidance law will impose a very complex motion on the particle, somewhat similar to a Brownian movement, so that the statistical distribution of possible positions represented by $|\Psi|^2$ seems to be rendered possible in this way.

However, as Takabayasi has pointed out, the very complicated motion of a particle that obeys such a rigorous law (the guidance formula) is not wholly comparable to a random motion such as the Brownian movement. One is thus led to surmise that the justification of the probability of presence given by $|\Psi|^2$, analogous to the ergodic theorem of Classical Statistical Mechanics, will require the introduction of rather delicate considerations. We will come back to this subject in Chapter XIII, which is devoted to a justification, in the general case of the causal theory, of the statistical meaning attributed to $|\Psi|^2$.

We may, however, be immediately permitted to make an interesting point in this connection. Suppose an H atom in an *s* state with wave function Ψ_{n00} and in which the electron is motionless. The slightest external perturbation will give Ψ the form $\Psi_{n00} + \delta\Psi$, and the presence of $\delta\Psi$ in this expression, by modifying the phase ever so slightly, will, according to the guidance formula, impart to the electron a violent movement of a Brownian character. If we assume that the probability of presence is given by $|\Psi|^2$, we will have in the perturbed state a statistical distribution of the electron's positions given by $|\Psi_{n00} + \delta\Psi|^2$, which will be extremely close to the distribution given by $|\Psi_{n00}|^2$. We realize, thus, that since the state Ψ_{n00} is in reality always subject to slight external perturbations, the electron is generally never really at rest, but, on the contrary, continually impelled by violent motions resulting from the slight perturbations; and we then have a better understanding of how a probability of presence quite obviously given by $|\Psi_{n00}|^2$ can actually hold. Similar considerations will be taken up again in Chapter XIII.[2]

2. Interference in the vicinity of a mirror (Wiener fringes)

In my book, *An Introduction to the Study of Wave Mechanics* [2], I studied the motion that photons must possess in the vicinity of a mirror according to the causal theory. I take up this theory again, considering the case of a particle of any non-zero mass ($m_0 \neq 0$) and obeying the Klein-Gordon equation.

a. A perfectly reflecting mirror

Let us assume a plane monochromatic Ψ wave striking a perfectly reflecting mirror at an angle of incidence ϑ.

Before entering the region of interference where the Wiener fringes appear, the plane wave has the initial form

$$\Psi_1 \sim e^{i\omega t} e^{-\frac{i\omega}{V}(x \sin\vartheta + z \cos\vartheta)} . \tag{14}$$

In the region of interference there is a superposition of the incident wave Ψ_1, and the reflected wave Ψ_2,

$$\Psi_2 \sim e^{i\omega t} e^{-\frac{i\omega}{V}(x \sin\vartheta - z \cos\vartheta)} e^{i\delta} \tag{15}$$

δ being the phase difference which may arise as a result of the reflection. One then has in the region of interference

$$\Psi = \Psi_1 + \Psi_2 = \sqrt{2} \cos\left(\frac{\omega}{V} z \cos\vartheta + \frac{\delta}{2}\right) e^{i\omega\left(t - \frac{x}{V}\sin\vartheta\right)} e^{i\frac{\delta}{2}} . \tag{16}$$

[2] For further remarks on the hydrogen atom according to the theory of the Double Solution, see ref. [17] pp. 72–73.

The Ψ wave is normalized, which introduces the factor $\sqrt{2}$. One thus has

$$
\begin{aligned}
a &= \sqrt{2}\cos\left(\frac{\omega}{V} z\cos\vartheta + \frac{\delta}{2}\right) \\
\varphi &= \hbar\omega\left(t - \frac{x}{V}\sin\vartheta\right) + \text{const.}
\end{aligned}
\tag{17}
$$

We thus have, within the region of Wiener fringes, according to the guidance formula

$$
\mathbf{v} = -c^2 \frac{\operatorname{grad}\varphi}{\dfrac{\partial\varphi}{\partial t}} \tag{18}
$$

whence

$$
v_y = v_z = 0 \qquad v_X = -c^2 \frac{\partial\varphi}{\partial x}\Big/\frac{\partial\varphi}{\partial t} = \frac{c^2}{V}\sin\vartheta = v_0\sin\vartheta \tag{19}
$$

$v_0 = c^2/V$ being the particle's velocity in the incident plane wave.

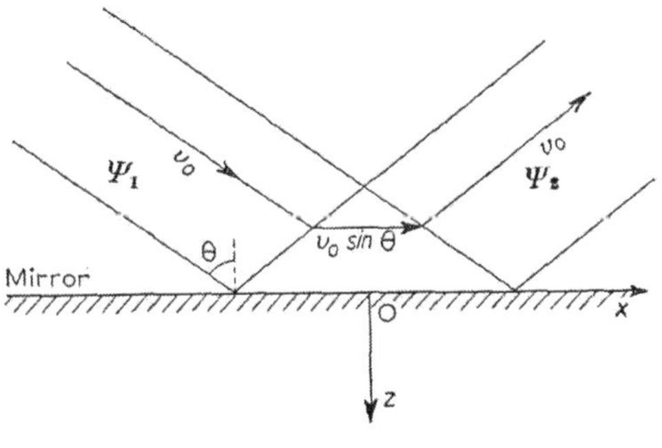

Fig. 7

The particle, which first possesses a uniform rectilinear motion with velocity v_0 in the direction of incidence, passes through the region of interference with a velocity $v_0 \sin\vartheta$ and then leaves it, moving rectilinearly in the direction of reflection with a velocity v_0. But it should be pointed out that on entering and on leaving the region of interference, the particle has a complicated motion because, along the boundaries of the incident beam and of the reflected beam, both of which are laterally limited, the representation of these beams by a plane monochromatic wave is no longer correct.

The quantity

$$a^2 = |\Psi|^2 = 2 \cos^2 \left(\frac{\omega}{V} z \cos \vartheta + \frac{\delta}{2} \right) = 1 + \cos \left(\frac{2\omega}{V} z \cos \vartheta + \delta \right) \quad (20)$$

gives the probability of the presence of particles as a function of the coördinate. We see that statistically the particles are distributed in layers separated by a distance of $(V/\omega)(\pi/\cos \vartheta) = \lambda/2 \cos \vartheta$ where the density is maximum, with equidistant layers where the density is zero. We do, in fact, once more have the classical Wiener fringes.

b. A partially reflecting mirror

Let us now assume that we are dealing with a partially reflecting mirror where the coefficient of reflection is η with $0 \leq \eta < 1$.

We will then have for the Ψ wave in the interference region

$$\Psi = \Psi_1 + \Psi_2 \sim e^{i\omega t} e^{-\frac{i\omega}{V} x \sin \vartheta} \left(e^{-i \frac{\omega}{V} z \cos \vartheta} + \eta e^{i \frac{\omega}{V} z \cos \vartheta} e^{i\delta} \right). \quad (21)$$

The calculation of $a^2 = |\Psi|^2$ easily gives

$$a^2 = \text{const.} \left[1 + \eta^2 + 2\eta \cos \left(\frac{2\omega}{V} z \cos \vartheta + \delta \right) \right]. \quad (22)$$

We again find bright fringes for the values of z which make the cosine equal to one, and dark fringes in between for the values of z which make the cosine equal to minus one; but in this case the dark fringes are not black, for the intensity in them is equal to $(1 - \eta)^2 > 0$.

To calculate the derivative of the phase $\partial \varphi / \partial q$ with respect to any of the spatial variables, q, we proceed from the formula

$$\frac{\partial \varphi}{\partial q} = \frac{\hbar}{2ia^2} \left(\Psi^* \frac{\partial \Psi}{\partial q} - \Psi \frac{\partial \Psi^*}{\partial q} \right) \quad (23)$$

which is immediately verifiable from the expression $\Psi = a e^{\frac{i}{\hbar}\varphi}$ with a and φ real. The guidance formula then gives us

$$\begin{aligned} v_x &= - c^2 \frac{\partial \varphi}{\partial x} \Big/ \frac{\partial \varphi}{\partial t} = \frac{c^2}{V} \sin \vartheta = v_0 \sin \vartheta, \qquad v_y = 0; \\ v_z &= - c^2 \frac{\partial \varphi}{\partial z} \Big/ \frac{\partial \varphi}{\partial t} = \frac{c^2}{V} \cos \vartheta \frac{1 - \eta^2}{a^2} = v_0 \cos \vartheta \frac{(1 - \eta^2)}{a^2} > 0. \end{aligned} \quad (24)$$

The particles, thus, approach the mirror in the region of interference, which means that some of them reach the semi-transparent mirror and succeed in passing into the transmitted beam, while others.

will not reach the mirror and will pass into the reflected beam. The proportion of each kind will, of course, depend on the value of η.

c. The appearance of velocities greater than the velocity of light in a vacuum

One of the consequences—and a seemingly surprising one—of formula (24) is that at certain points in the field of interference the velocity of the particle may be greater than c.

Indeed, we find

$$v^2 = v_x^2 + v_z^2 = \frac{c^4}{V^2}\left[\sin^2\vartheta + \cos^2\vartheta\,\frac{(1-\eta^2)^2}{a^4}\right]. \tag{25}$$

Now, in the dark fringes,

$$\cos\left(\frac{2\omega}{V}z\cos\vartheta + \delta\right) = -1 \qquad \text{and } a^2 = (1-\eta)^2$$

whence

$$v^2 = \frac{c^4}{V^2}\left[\sin^2\vartheta + \cos^2\vartheta\,\frac{(1+\eta)^2}{(1-\eta)^2}\right] \tag{26}$$

a quantity that may be greater than c^2 if $c^4/V^2 = v_0^2$ is sufficiently close to c^2 and if $(1+\eta)/(1-\eta)$, always greater than 1, is sufficiently large.

To present the question more clearly, let us consider the simple case of normal incidence where $\cos\vartheta = 1$ and $\sin\vartheta = 0$. If the mirror is completely reflecting ($\eta = 1$), in that case we have $v_x = v_y = v_z = 0$; and the particles are motionless in the Wiener fringes. One might be surprised at this result alone, for one might wonder how the particles, if they are immobile in the region of interference, can somehow reach the bright fringes. This objection may once again be overcome by noting that the plane monochromatic wave on which we have based our reasoning has no existence as a physical reality and that one must always consider limited wave-trains. For this reason, our theory describes only the *steady state* realized while a train of Ψ waves strikes the mirror and not the *transient state* which exists when the wave-front strikes the mirror—a state in which the particles have a complicated motion during the course of which they come to occupy the bright fringes.

Still considering normal incidence, let us return to the study of the partially reflecting mirror. We then have

$$v_x = v_y = 0 \qquad v_z = \frac{c^2}{V}\,\frac{1-\eta^2}{a^2} \qquad v^2 = \frac{c^4}{V^2}\left(\frac{1-\eta^2}{a^2}\right)^2 \tag{27}$$

and v will be greater than c if

$$\frac{c^2}{V^2}\left(\frac{1-\eta^2}{a^2}\right)^2 > 1. \tag{28}$$

If we put $u = 2\omega/V + \delta$, which gives $a^2 = 1 + \eta^2 + 2\eta \cos u$, condition (28) will be written

$$1 - \eta^2 > \frac{V}{c}(1 + \eta^2 + 2\eta \cos u). \tag{29}$$

It is easily seen that this condition may also be written

$$\frac{\frac{c}{V}(1-\eta^2) - (1+\eta^2)}{2\eta} > \cos u \geqq -1. \tag{30}$$

For η sufficiently close to unity, there is thus a small region in the vicinity of each dark fringe where, when the foregoing condition obtains, one has $v > c$. This region shrinks as η increases and is reduced to the center line of the dark fringe itself in the case of a completely reflecting mirror ($\eta=1$). In the latter case, there could be no velocities greater than c except on this central line, but since there is no particle in the central line, no velocity higher than c actually does appear. This observation makes it possible to restore the continuity between the formulas obtained here and those given in section *a* for the completely reflecting mirror.

When there are layers in the region of interference where the inequalities (30) are satisfied, the velocity of the particle is, according to the guidance formula, greater than c in these layers. Now we know that we must have

$$W = \frac{\partial \varphi}{\partial t} = \frac{M_0 c^2}{\sqrt{1-\beta^2}} = h\nu = \hbar\omega \tag{31}$$

M_0 being the variable rest-mass of the particle in the field of interference. Since $\hbar\omega$ is a real quantity and since $\sqrt{1-\beta^2}$ becomes imaginary for $v > c$, we may conclude that in every place where v is greater than c, M_0 must be imaginary. We are going to verify this by starting with the definition of M_0.

$$M_0^2 = m_0^2 + \frac{\hbar^2}{c^2}\frac{\Box a}{a} \tag{32}$$

which gives us here

$$M_0^2 c^2 = m_0^2 c^2 - \hbar^2 \frac{\Delta a}{a} = m_0^2 c^2 - \frac{\hbar^2}{4a^2}\left[2a^2 \Delta a^2 - (\text{grad } a^2)^2\right] \quad (33)$$

as is easily verifiable. Since $a^2 = 1 + \eta^2 + 2\eta \cos u$, this gives us

$$M_0^2 c^2 = m_0^2 c^2 + \frac{\hbar^2}{a^4}\left[\eta^2 k^2 \sin^2 u + \eta a^2 k^2 \cos u\right] \quad (34)$$

with

$$k = \frac{2\omega}{V} = \frac{4\pi}{\lambda}.$$

Now from $\lambda\nu = V$, $W = h\nu$, $\lambda = h/p$, we easily obtain $W^2/c^2p^2 = V^2/c^2$ and, since we have $W^2/c^2 = p^2 + m_0^2 c^2$, we may deduce

$$4\frac{m_0^2 c^2}{\hbar^2} = k^2 \frac{1 - \frac{c^2}{V^2}}{\frac{c^2}{V^2}}. \quad (35)$$

Beginning with (34), we then find

$$M_0^2 = \frac{\hbar^2 k^2}{4a^4}\left(4\eta^2 \sin^2 u + 4a^2 \eta \cos u + a^4 \frac{1 - \frac{c^2}{V^2}}{\frac{c^2}{V^2}}\right). \quad (36)$$

M_0 will then be imaginary if

$$\eta^2 \sin^2 u + a^2 \eta \cos u + \frac{a^4}{4}\frac{1 - \frac{c^2}{V^2}}{\frac{c^2}{V^2}} < 0. \quad (37)$$

By replacing $\sin^2 u$ by $1 - \cos^2 u$ and by expanding the expressions for a^2 and a^4, we obtain

$$\eta \cos^2 u + (1 + \eta^2) \cos u + \eta \frac{c^2}{V^2} + \frac{(1 + \eta^2)^2}{4\eta}\left(1 - \frac{c^2}{V^2}\right) < 0. \quad (38)$$

Now the second degree polynomial in $\cos u$ which forms the left-hand side of the inequality (38) has the two roots $[-(1 + \eta^2) \pm (c/V)(1 - \eta^2)]/2\eta$. Since $0 \leqq \eta < 1$ and $0 \leqq c/V \leqq 1$, we see that the inequality (38) gives

$$\frac{\frac{c}{V}(1-\eta^2)-(1+\eta^2)}{2\eta} > \cos u \geqq -1. \tag{39}$$

And one is back to inequality (30). M_0, which reduces to zero for $v = c$, is thus imaginary for $v > c$.

So at last we have established the following result: For every particle obeying the Klein-Gordon equation, there arises as a result of the guidance formula, a situation, in the neighborhood of a partially reflecting mirror, wherein the velocity of the particle can at certain points be greater than c.[3] Is not this conclusion in outright contradiction to the Theory of Relativity?

It must first of all be pointed out that, if Relativistic Dynamics rejects the possibility of a particle velocity greater than c, that is because it takes for granted the formula

$$W = \frac{m_0 c^2}{\sqrt{1-\beta^2}} \tag{40}$$

where m_0 is a constant. From it there follows that, if v approaches c, W approaches infinity. Thus it would be necessary to supply infinite energy to the particle in order to impart to it the velocity c. Moreover, for $v > c$, W would be imaginary.

But the causal theory introduces a reaction of the wave upon the particle that is completely unknown in the older Relativistic Dynamics. This reaction is expressed in the equations of motion by the introduction of the quantum potential Q, and we have shown that there results from this a dynamics of the particle in which the rest-mass M_0 is in general variable according to the law

$$M_0 = \sqrt{m_0^2 + \frac{\hbar^2}{c^2}\left(\frac{\Box a}{a}\right)} \tag{41}$$

[3] The theory here developed, since it is valid only for particles of zero spin, is not applicable to photons and to actual Wiener optical fringes. The Wave Mechanics we have set forth in other works shows clearly that photons are particles with spin one, whose wave equations are Maxwell's equations. So it is evident that from what will be said further on—especially in the appendix, the motion of photons must be defined not by the guidance formula as presented in the above, but rather as it will be modified by the "lines of current" of the associated electromagnetic wave. This is tantamount to saying that the motion of photons must be defined with the help of a Poynting vector. If the preceding theory is reconsidered in that way, it seems to me certain that velocities greater than c could no longer appear.

so that the expression for the energy becomes

$$W = \frac{M_0 c^2}{\sqrt{1 - \beta^2}}. \tag{42}$$

Energy does not, thus, become infinite for $v = c$, nor imaginary for $v > c$. It thus seems that the reason which, in regular Relativistic Dynamics, renders impossible any velocity v greater than c, here no longer exists.

But, one may say, Einstein's analyses of the concepts of space and time, and of the measurements of length and duration, which were the very bases of the development of the Theory of Relativity, rest essentially on the postulate that no signal can be transmitted with a velocity v greater than c, the velocity of light in a vacuum. Does not the contention that an energy-transporting particle could, in certain cases, move with a velocity greater than c go counter to this fundamental postulate?

It seems that even this objection may be obviated. Particle velocities higher than c seem only to appear in the causal theory in certain regions of interference-fields. Now the appearance of interferences is brought about by well defined experimental apparatus with well determined initial and limiting conditions (in the case of Wiener fringes, incident Ψ wave-trains that are almost monochromatic and limiting conditions on the surface of the mirror). It is thus impossible to show the presence of the particle at two nearby points M and M′ of the interference-field at the successive instants t and t' in such a way that it will play the role of a signal with a velocity greater than c. To do this it would be necessary to observe the localization of the particle at M and M′ at the times t and t', and this would require another kind of experimental device—one incompatible with obtaining Wiener fringes. The similarity between this argument and certain of Bohr's reasonings will be noted, but here the argument is utilized within the framework of the causal interpretation without giving up the possibility of localizing the particle.

It must, indeed, be noted that what changes the particle's motion when the experimental apparatus is modified is the fact that, from the point of view of the causal theory, the new apparatus introduces new limiting conditions for the u waves—the only physical realities; and thus a correlative adjustment of the fictitious Ψ wave that is to be associated with the u wave must be made. In short, particle velocities greater than c, which the guidance formula indicates for certain cases, cannot, it seems, be actually revealed; and thus they cannot be used for signals of a velocity greater than c.

In the region of space-time where $v > c$, the world line of the particle is space-like; its frequency and its proper time are imaginary, at least if we still assume the usual definitions; but the proper phase $\nu_0 t_0$ remains real. This conception may appear strange, but is it any more so than Feynman's view that the positron is an electron that moves backward in time? There are, moreover, possible points of similarity between this idea of Feynman's and the question here being studied. Indeed, in order for the world line of an electron to reverse its direction in time, that line must exhibit either a vertex or elements of a space-like character. In both cases, action and phase will always vary in the same way all along the world line.[4]

3. A recent objection of Einstein to the guidance formula

In his recent contribution to the anniversary book dedicated to Max Born, Einstein, while continuing to press for the reestablishment of a causal interpretation in Wave Mechanics, raised an objection against the guidance formula.

Einstein proceeds from the principle that, every time we are concerned with a macroscopic body, we must end up with the images furnished by Classical Mechanics, which then gives us beyond any doubt a description that is very close to actual reality. Having laid down this principle, he considers the following problem: Let us consider a particle that moves along an axis x and then rebounds between two mirrors placed normally on the x-axis at the points $x = 0$ and $x = l$. Standard Wave Mechanics associates with this particle's motion, if it has a well determined energy, a stationary Ψ wave with $\Psi = 0$ at $x = 0$ and $x = l$ and of the form

$$\Psi = a_n \sin \frac{n\pi x}{l} e^{\frac{i}{\hbar} E_n t} \qquad \text{with } E_n = \frac{n^2 h^2}{8ml^2} \tag{43}$$

with n an integer. In the usual interpretation this wave represents the possibility of two motions of equal momentum $p = nh/2l$, one taking place from right to left and the other from left to right, with the two

[4] But all velocities greater than c would perhaps disappear if we systematically defined the guidance by lines of current. (See Appendix).

motions having the same probability. From the point of view of macroscopic physics, *if the particle is macroscopic*, it has one of these two motions to the exclusion of the other. The usual interpretation, says Einstein, thus represents precisely the statistical situation in the course of time, but not the real instantaneous state of the macroscopic particle. It is thus an exact statistical interpretation, but "incomplete" as a description of physical reality.

Such is Einstein's point of view, and it strikes us as thoroughly defensible. But now we come to the objection, based upon this view, which he brought against the guidance formula. In formula (43), the phase φ of the Ψ wave is reduced to $E_n t$ and is independent of x; the guidance formula thus gives us $v = -1/m \text{ grad } \varphi = 0$. The particle would be immobile, and if the particle has a macroscopic mass and constitutes a small solid sphere in the usual sense of the term, we find ourselves in contradiction to macroscopic physics which affirms—and certainly with reason—that the small solid sphere must possess a back-and-forth motion along Ox with an alternate rebound from each of the two mirrors. Einstein concludes from this that the guidance formula as well cannot represent physical reality.

Einstein's considerations are, on the whole, very interesting; but they call immediately for one reservation: One may, in fact, point out that the particle, if it is macroscopic, is necessarily formed of a collection of numerous elementary particles and that the Ψ wave is then associated with the center of gravity of the system, and this makes the interpretation more difficult. But, apart from this reservation, one can, it seems to me, find two very suggestive answers to Einstein's objection to the guidance formula.

The first answer arises from the following observation: In order for the form of the Ψ wave adopted in Einstein's example to be considered valid, it is necessary that the obstacles limiting the particle's motion and the propagation of its associated wave may be considered as plane mirrors for the Ψ wave. Now, these mirrors are necessarily formed of atoms in thermal motion, and as a consequence the precision with which their surfaces may be defined cannot go beyond some fraction of an Ångström unit. Using as a guide a theory developed by Debye for the evaluation of the influence of thermal motions of atoms in a crystal on the diffraction of X-rays, one can see that the wave-length must not be greater than approximately 10^{-10} cm. The validity of the expression adopted for Ψ thus requires that the condition

$$\lambda = \frac{h}{mv} > 10^{-10}\ \text{cm} \tag{44}$$

be fulfilled. The expression (44) shows that if the particle has a macroscopic mass (let us say, greater than 10^{-9} g), the velocity v must be very nearly zero. Then in order that expression (43) of Ψ may be considered valid for a particle of macroscopic mass, the particle's velocity must be approximately zero, and then the value $v = 0$ supplied by the guidance formula for the velocity is approximately true. It does, therefore, seem that Einstein's objection to the guidance formula may be answered in this way. Let us also point out that we here meet with the idea that one must attribute great importance to the possible fluctuations of limiting conditions, and we will find this idea appearing again when we discuss the justification of the statistical meaning attributed to $|\Psi|^2$ by the theory of the Double Solution.[5]

But let us go on to the second way in which Einstein's objection may be answered. It arises from the observation that the Ψ wave of a particle must always be considered as forming a wave-train of limited dimensions. One can then, it seems, still admit that a Ψ wave-train has limited dimensions which cannot exceed a certain multiple (very large) of the wave-length. It is in this way that, in the case of a photon, we know that the wave-train cannot exceed a few million wave-lengths. A similar limitation must be valid for the other sorts of particles. Since, for a given energy, the wave-length $\lambda = h/\sqrt{2m\text{E}}$ diminishes as the mass increases, we see that the wave-train associated with a particle of discernible energy will finally, as the mass increases, have a length much smaller than the distance l separating the mirrors in Einstein's example. For a sufficiently great mass, it will thus no longer be possible to imagine a stationary wave resulting from the superposition, between the two mirrors, of two waves propagated in opposite directions. One will, on the contrary, have to imagine a train of Ψ waves of small dimensions reflecting alternately on each of the two mirrors, and that image will correspond exactly to the macroscopic image of a tiny sphere oscillating along a straight line and alternately striking and

[5] Note that, for electrons ($m \simeq 10^{-27}$ g), v may attain values very close to c and that, for molecules ($m \lesssim 10^{-24}$ g) one may reach velocities of the order of 10^6 cm/sec = 10 km/sec, without in any way affecting the validity of inequality (44). One can thus consider the walls of an enclosure well defined as mirrors for the Ψ wave for very rapid electronic motions ($v \simeq c$) and for molecular motions in a gas under normal conditions —which in both cases justifies the use of expression (43) for the Ψ wave.

rebounding from the two parallel walls. This very interesting answer to Einstein's objection to the guidance formula illustrates once again the importance of the fact that every wave-train has limited dimensions.

Chapter XII

THE TRANSITION FROM SINGLE-PARTICLE WAVE MECHANICS TO THE WAVE MECHANICS OF PARTICLE SYSTEMS

1. The nature of the problems in the causal theory

We have seen how Schrödinger succeeded in constructing the Wave Mechanics of an ensemble of particles by associating with this system's motion the propagation of a wave in a corresponding configuration space. But one obtains in this way an essentially non-relativistic theory, because it expresses the interactions between particles by functions of the distances separating them at the same instant, and because this assumption of the instantaneous propagation of interactions is contrary to the relativistic principle that every perturbation is propagated in space with a finite velocity.

In spite of the great formal beauty of Schrödinger's theory and the remarkable successes that it immediately achieved, it was possible to raise serious objections to it from the very outset.

In the first place, just as Bohr did in his approach, Schrödinger identified particles with continuous wave-trains, and thus no longer considered particles as point-like. Consequently, it seems contradictory to assign well defined coördinates to them. And it therefore seems unjustifiable to consider a configuration space formed of 3N coördinates of the constituents of the system.

Secondly, configuration space is obviously an abstract, fictitious space devoid of physical reality. By considering the Ψ wave as being propagated in that space one necessarily deprives the Ψ wave of all physical reality. If the waves of Wave Mechanics were to retain any physical reality, it seemed to me that we had to be able to consider the motion of the particles, as well as the evolution of the wave phenomenon associated with them, in the framework of three-dimensional physical space. But Schrödinger's method necessarily implies the use of configuration space and no longer permits us to represent the physical phenomenon of the motion of the particles in the framework of physical space. Of course, Classical Mechanics likewise often utilized configuration space, but it did not *have* to; it could reason by considering the motion of the point-masses of the system in three-dimensional space, and it utilized configuration space only as a mathematical artifice that made it possible to present certain calculations more elegantly or perform them more easily. From the very moment that Schrödinger's Papers appeared, while I recognized the correctness of the results obtained by his method, I was struck by the paradoxical character of the very principle on which the method is based.

Viewing the matter from the point of view of the Double Solution, it seemed to me necessary to rethink the whole question in a different way. For me the particles were embedded in a wave phenomenon in which they constituted a singularity (today I would say, "a small singular region"). Each of the singular regions was to be considered as a center of force conditioning the propagation of the wave phenomena associated with the other particles and, consequently, on the motion of the other particles. From this there must result a complicated motion of the ensemble of the "singular-region particles". One can obviously represent this motion by the displacement of a representative point in a configuration space formed by the coördinates $x_1 \ldots z_N$ of N particles. There is no difficulty here, since the particles which are very small singular regions of almost point-like character have at practically every instant a well-defined position and well-defined coördinates. One might then think that by defining the propagation of a *purely fictitious* wave $\Psi(x_1 \ldots z_N, t)$ in configuration space, we could make the quantity $|\Psi|^2$ play the role of a probability of the presence of a representative point at the various points in configuration space—which would once again result in the Schrödinger theory.

One must realize that, if this idea is correct, the description obtained by the use of the propagation of the continuous Ψ wave in the system's configuration space is much less complete than the description furnished by the theory of the Double Solution, which considers the u waves of N particles in physical space with their mobile singular regions. One would obtain by this latter description a picture that would be correct in the physical space not only of the N particles, but also of the N wave phenomena in which they are embedded; the configuration-space method, on the other hand, would only supply statistical information concerning the displacement of the N singular regions while totally omitting the extended wave phenomena of which they are the center. In this way Schrödinger's method would be justified and its success explained; but, at the same time, one would see that it conceals, so to speak, a certain number of much more complex wave phenomena which take place in three-dimensional physical space.

Such were the ideas which guided me in 1927 when I drew up the section of my paper on the Double Solution devoted to the Wave Mechanics of systems. I will first reproduce my 1927 arguments, and then add the results of recent research. I will, moreover, always consider, for the sake of simplicity, the case of a system made up of two particles, because the step from the case of two particles to that of N particles does not entail any difficulty of principle.[1]

2. The arguments of my 1927 paper

Let us consider two particles, each of which constitutes a small singular region in their respective u waves, and let us write the two equations of propagation for the u waves, *assuming that no external action is exerted on the system*

$$\begin{aligned} \Box u_1 + \frac{2i}{\hbar}\frac{F_{12}}{c^2}\frac{\partial u_1}{\partial t} + \frac{1}{\hbar^2}\left(m_0^2 c^2 - \frac{F_{12}^2}{c^2}\right) u_1 = 0 \\ \Box u_2 + \frac{2i}{\hbar}\frac{F_{21}}{c^2}\frac{\partial u_2}{\partial t} + \frac{1}{\hbar^2}\left(m_0^2 c^2 - \frac{F_{21}^2}{c^2}\right) u_2 = 0 \end{aligned} \tag{1}$$

m_1 and m_2 being the *rest* masses of the two particles, F_{12} and F_{21} representing the action exerted on each of the particles by the presence of the other. I put

$$\begin{aligned} F_{12} = F[\sqrt{(x - x_2)^2 + (y - y_2)^2 + (z - z_2)^2}]; \\ F_{21} = F[\sqrt{(x - x_1)^2 + (y - y_1)^2 + (z - z_1)^2}]. \end{aligned} \tag{2}$$

Here the coordinates x, y, z that appear in the operators $\Box$ are the regular coordinates of three-dimensional physical space. Expressions (2) indicate that the value of F_{12} at the point occupied by the first particle is the same as that of F_{21} at the point occupied by the second particle. If r designates the distance between the two particles, this common value is $F(r)$, which is in keeping with the classical principle of action and reaction. But it should be noted that F_{12} and F_{21} are defined at every instant at every point in physical space. The propagation in space of each of the two waves u_1 and u_2 is thus found to depend, at

[1] However, see end of Section 4.

every point, upon the instantaneous value of the potential corresponding to the simultaneous position of the singular region in the other wave.

One is forced to assume the existence of a solution for the two equations (1) which describes a mobile singular region, and, in order to compare it with Schrödinger's theory we will always have to be satisfied with the Newtonian approximation.

In Classical Mechanics there exists for this two-particle system a Jacobi function $S(x_1, \ldots, z_N, t)$ such that

$$m_1 v_{1x} = -\frac{\partial S}{\partial x_1}, \ldots \qquad m_2 v_{2z} = -\frac{\partial S}{\partial z_2}. \tag{3}$$

Can we define a function $\varphi(x_1, \ldots, z_N, t)$ which will play the role of S?

For a moment we assume that the motion of the second particle is known to us, and then the motion of the first one will take place in a field that will be a known function of x, y, z, t—a case we have already studied. So we know that we can associate with the collection of possible motions of the first particle the propagation, in three-dimensional space, of a continuous wave

$$\Psi_1 = a_1(x, y, z, t) e^{\frac{i}{\hbar}\varphi(x, y, z, t)}.$$

Now the theories developed for a single particle in a given field make it possible to express that particle's motion by means of Lagrange equations of the form

$$\frac{d}{dt}\left(\frac{\partial \mathscr{L}_1}{\partial v_{1x}}\right) = \frac{\partial \mathscr{L}_1}{\partial x}, \ldots \tag{4}$$

with

$$\mathscr{L}_1 = \tfrac{1}{2} m_1 v_1^2 - F(r) - \mathcal{Q}_1 \tag{5}$$

where

$$\mathcal{Q}_1 = \frac{\hbar^2}{2m_1}\left(\frac{\Box a_1}{a_1}\right) \simeq -\frac{\hbar^2}{2m_1}\left(\frac{\Delta a_1}{a_1}\right). \tag{6}$$

Similarly, by assuming for a moment that we know the motion of the first particle, the equations of the second particle may be determined by writing the Lagrange equations

$$\frac{d}{dt}\left(\frac{\partial \mathscr{L}_2}{\partial v_{2x}}\right) = \frac{\partial \mathscr{L}_2}{\partial x_2} \tag{7}$$

with

$$\mathscr{L}_2 = \tfrac{1}{2}m_2 v_2^2 - F(r) - Q_2, \tag{8}$$

where

$$Q_2 = \frac{\hbar^2}{2m_2} \frac{\Box a_2}{a_2} \simeq -\frac{\hbar^2}{2m_2}\left(\frac{\Delta a_2}{a_2}\right). \tag{9}$$

Now we must solve the two groups of Lagrange equations (4) and (7) *simultaneously*. In Classical Mechanics where one may consider the individual Lagrange equations with Q_1 and Q_2 equal to zero, it is easily seen that one can define a Lagrange function for the entire system $\mathscr{L}(x_1, y_1, z_1; x_2, y_2, z_2; v_{1x}, v_{1y}, v_{1z}; v_{2x}, v_{2y}, v_{2z})$ such that the system's motion is given by the six equations

$$\frac{d}{dt}\left(\frac{\partial \mathscr{L}}{\partial \dot{q}}\right) = \frac{\partial \mathscr{L}}{\partial q}\left(q = x_1 \ldots z_2,\ \dot{q} = \frac{dq}{dt}\right). \tag{10}$$

For this, one has only to put

$$\mathscr{L} = \tfrac{1}{2}m_1 v_1^2 + \tfrac{1}{2}m_2 v_2^2 - F(r) = \mathscr{L}_1 + \mathscr{L}_2 + F(r) \tag{10'}$$

that is, to take for $\mathscr{L}$ the sum of the kinetic energy of $\mathscr{L}_1$, and $\mathscr{L}_2$ minus *one-half the sum* of the potential terms. From the Lagrangian, moreover, for the energies E_1 and E_2 of the two particles and for the total energy E of the system, one easily obtains the expressions

$$E_1 = \tfrac{1}{2}m_1 v_1^2 + F(r)$$

$$E = \tfrac{1}{2}m_1 v_1^2 + \tfrac{1}{2}m_2 v_2^2 + F(r) = E_1 + E_2 - F(r) \tag{11}$$

$$E_2 = \tfrac{1}{2}m_2 v_2^2 + F(r)$$

These show that the total energy is the sum of the individual kinetic energies plus *one-half the sum* of the individual potential energies.

The fact that $\mathscr{L}$ is not the sum of $\mathscr{L}_1$ and $\mathscr{L}_2$ and that E is not the sum of E_1 and E_2 (because the potential-energy term appears only once in the Lagrange function and in the total energy of the system) is not, so far as I know, specifically pointed out in treatises on Mechanics. This fact is the necessary consequence of the principle of action and reaction, and it symbolizes a sharing of the potential energy by the two particles.

In order to be able to apply the same formalism in the theory of the Double Solution, it is necessary that the terms Q_1 and Q_2 exhibit the same character of mutual action as do the terms F_{12} and F_{21}. In my 1927 paper, I assumed that this was the case, and that was clearly a weak point in my argument, because it should have been fully demonstrated that such was indeed the case.

If this hypothesis is admitted, it is easily seen that we can obtain a Lagrange function for the two-particle system by putting

$$\mathscr{L} = \tfrac{1}{2}m_1 v_1^2 + \tfrac{1}{2}m_2 v_2^2 - F(r) - Q(r) \tag{12}$$

where $Q(r)$ is the assumed common value $Q_1(x, y, z, t)$ at this point occupied by the second particle and of $Q_2(x, y, z, t)$ at the point occupied by the first particle, at the same instant t. The motion of the two particles must then be given by the Lagrange equations

$$\frac{d}{dt}\left(\frac{\partial \mathscr{L}}{\partial \dot{q}}\right) = \frac{\partial \mathscr{L}}{\partial q} \qquad (q = x_1 \ldots z_N). \tag{13}$$

Now, in the Classical Mechanics of systems it is demonstrated that, when the Lagrangian approach is valid, one can find a function $S(x_1, \ldots, z_N, t)$ of the variables of space and time such that the components of the momentum are given by relations (3). Since we here find a Lagrangian description for the motion of two particles, it must be possible to define a function $\varphi(x_1, \ldots, z_2, t)$ of the space variables and time such that one will have

$$m_1 v_{1x} = -\frac{\partial \varphi}{\partial x_1} \ldots m_2 v_{2z} = -\frac{\partial \varphi}{\partial z_2}. \tag{14}$$

Taking Schrödinger's method as our model, we will neglect for the time being the u_1 and u_2 waves, which develop in three-dimensional physical space, and concentrate our attention solely on the successive positions of the two singular regions. In this way we will obtain an incomplete theory which, by its very nature, allows an important part of physical reality to elude our grasp, but which—as the success of Schrödinger's theory proves to us—may permit us to find once more the statistics of the successive positions of the particles represented by the motion of the representative point of the system in the fictitious space formed by the six coördinates $x_1, \ldots, z_2$ of the two singular regions.

Following the same line of reasoning that we did in the case of the Ψ wave of a single particle, we will assume that at the initial instant the two particles are far enough apart that their interactions are negligible and that their respective waves are wave-trains that are essentially monochromatic. The initial velocities of the particles are thus known, but their respective initial positions in their wave-train are unknown. There will correspond to the various hypotheses that one can make about the initial positions, and which it is natural to consider as equally probable (providing, as we have seen, that the necessary justification is made), various trajectories of the representative point in configuration space. The motion of the probability density is stationary and obeys the continuity equation

$$\operatorname{div}(\rho \mathbf{v}) = 0 \tag{15}$$

where $\rho(x_1, \ldots, z_2)$ is the probability density and $\mathbf{v}$ its six-component velocity in configuration space.

If one takes formulas (14) into account, which give the velocity components of the particles in terms of φ, it is seen that equation (15) will be written

$$\sum_{x,y,z}\left[\frac{1}{m_1}\frac{\partial\varphi}{\partial x_1}\frac{\partial \ln\rho}{\partial x_1} + \frac{1}{m_2}\frac{\partial\varphi}{\partial x_2}\frac{\partial}{\partial x_2}\ln\rho\right] + \frac{1}{m_1}\Delta_1\varphi_1 + \frac{1}{m_2}\Delta_2\varphi_2 = 0. \tag{16}$$

Now, if, with Schrödinger, we define the wave function $\Psi(x_1, \ldots, z_2, t)$ as a solution of the equation of propagation

$$\sum_{i=1}^{2}\frac{1}{m_i}\Delta_i\Psi + \frac{2}{\hbar^2}[\mathrm{E} - \mathrm{F}(r)]\Psi = 0 \tag{17}$$

in configuration space; and if we put

$$\Psi(x_1 \ldots z_2, t) = a(x_1 \ldots z_2)e^{\frac{i}{\hbar}\varphi(x_1,\ldots z_2,t)} \tag{18}$$

we find, after substitution in (17),

$$\sum_{xyz}\left[\frac{1}{m_1}\frac{\partial\varphi}{\partial x_1}\frac{\partial}{\partial x_1}\ln a^2 + \frac{1}{m_2}\frac{\partial\varphi}{\partial x_2}\frac{\partial}{\partial x_2}\ln a^2\right] + \frac{1}{m_1}\Delta_1\varphi + \frac{1}{m_2}\Delta_2\varphi = 0. \tag{19}$$

The comparison of (16) and (19) shows that the amplitude a of the fictitious Ψ wave in configuration space will, from a statistical point of view, play the same role for the representative point as theamplitude of the continuous Ψ wave does in the case of a single particle. In other words, keeping always in mind the hypothesis that all the positions of the two particles are equally probable in the isolated and almost monochromatic wave-trains that are associated with them in the initial state, the product $a^2 d\tau = |\Psi|^2 d\tau$ at every point in configuration space can be said to give the probability that the representative point is present at a given instant in the element $d\tau$; it will even give the probability in absolute value if we take care always to normalize the Ψ wave function in configuration space by use of the standard formula $\int d\tau|\Psi|^2 = 1$.

Such were the considerations that I developed in my 1927 paper. They had the advantage of bringing out clearly the two following points:

1. The use of configuration space is natural in the theory of the Double Solution because the particles at every instant have a well defined position in three-dimensional physical space, permitting us to define the configuration variables clearly—variables which have no such clear-cut meaning if we deny the existence of a particle's position at every instant and thus render the use of configuration space very difficult to justify.

2. The real evolution of the overall wave phenomena of the N particles must, from the point of view of the Double Solution, be described by the propagation in three-dimensional physical space of N u-waves each involving an almost point-like singular region; but, if one temporarily neglects the extended u waves in order to concentrate exclusively on the motion of the singular regions, one *can* bring in configuration space to represent, following Schrödinger's method, the probabilities of presence by means of the propagation of a Ψ wave —a purely fictitious wave, as are all the Ψ waves in the theory of the Double Solution. What ever may be the future fate of the theory of the Double Solution, these particular considerations, we feel, will remain of interest. In particular, they clearly show that, in passing from a description in physical space to a description in configuration space, we inevitably destroy the possibility of describing "fields" extended in physical space, since every point in configuration space can represent only the ensemble of certain singular points in these fields.

But the reasoning summarized in the beginning of this section has weak points. First of all, it assumes the hypothesis that quantum potentials have the same "mutual" character as ordinary potentials, without in any way justifying this hypothesis. In addition, we have implicitly assumed: 1) that there are no external fields acting on the particles; 2) that there do not exist, in the regions of space where the u waves develop, any obstacles causing interference or diffraction phenomena entailing the introduction of quantum potentials which would most certainly *not* be of mutual character. In other words, we have assumed that the potentials F and Q were solely dependent on the mutual actions of the two particles.

In order to overcome these restrictions, we have recently developed another way of attacking the question. We now explain this method.

3. Another approach to the question

In order to develop the new point of view, we will first of all demonstrate the following lemma.

Lemma

Let there be two variables x_1 and x_2 and a certain function F *of x_1 and x_2. Let us consider three functions $F_1(x_1, r)$, $F_2(x_2, r)$ and $F(x_1, x_2, r)$, and let us assume that we have the following relations between these functions:*

$$\begin{aligned} \left(\frac{\partial F}{\partial x_1}\right)_{x_2} &= \frac{\partial F}{\partial x_1} + \frac{\partial F}{\partial r}\frac{\partial r}{\partial x_1} = \left(\frac{\partial F_1}{\partial x_1}\right)_{x_2} = \frac{\partial F_1}{\partial x_1} + \frac{\partial F_1}{\partial r}\frac{\partial r}{\partial x_1} \\ \left(\frac{\partial F}{\partial x_2}\right)_{x_1} &= \frac{\partial F}{\partial x_2} + \frac{\partial F}{\partial r}\frac{\partial r}{\partial x_2} = \left(\frac{\partial F_2}{\partial x_2}\right)_{x_1} = \frac{\partial F_2}{\partial x_2} + \frac{\partial F_2}{\partial r}\frac{\partial r}{\partial x_2} \end{aligned} \tag{20}$$

Then the function F can be written in the form:

$$F(x_1, x_2, r) = F_{11}(x_1) + F_{22}(x_2) + F_{12}(r).$$

From the first relation (20), we deduce that

$$F(x_1, x_2, r) = F_1(x_1, r) + H(x_2),$$

and from the second relation (20) we likewise deduce

$$F(x_1, x_2, r) = F_2(x_2, r) + G(x_1).$$

Now that can be true only when we have

$$F_1(x_1, r) = F_{11}(x_1) + F_{12}(r); \qquad F_2(x_2, r) = F_{22}(x_2) + F_{21}(r) \tag{21}$$

with

$$F_{11} = G \qquad F_{22} = H \qquad \text{and } F_{12} = F_{21}.$$

From this we conclude, thus, that the hypotheses

$$\left(\frac{\partial F}{\partial x_1}\right)_{x_2} = \left(\frac{\partial F_1}{\partial x_1}\right)_{x_2} \quad \text{and} \quad \left(\frac{\partial F}{\partial x_2}\right)_{x_1} = \left(\frac{\partial F_2}{\partial x_2}\right)_{x_1}$$

impose on F the form

$$F(x_1, x_2, r) = F_{11}(x_1) + F_{22}(x_2) + F_{12}(r) \tag{22}$$

which proves the lemma.

Let us go back to the two-particle system we previously considered. Every point in physical space will be determined by the radius vector

R from the origin to the given point. The positions of the two particles at the instant t will thus be defined by the two vectors $\mathbf{R}_1(t)$ and $\mathbf{R}_2(t)$. The position of a point **R** in space in relation to the first particle will be defined by the vector $\mathbf{r}_1 = \mathbf{R} - \mathbf{R}_1(t)$; its position in relation to the second particle will be defined by the vector $\mathbf{r}_2 = \mathbf{R} - \mathbf{R}_2(t)$; finally, the position of the first particle in relation to the second will be given by the vector $\mathbf{r}_{12} = \mathbf{R}_1(t) - \mathbf{R}_2(t)$. The corresponding distances will have as their expressions

$$|\mathbf{r}_1| = \sqrt{|\mathbf{R} - \mathbf{R}_1(t)|^2}; \qquad |\mathbf{r}_2| = \sqrt{|\mathbf{R} - \mathbf{R}_2(t)|^2} \tag{23}$$
$$r_{12} = \sqrt{|\mathbf{R}_1(t) - \mathbf{R}_2(t)|^2}.$$

Having established this, we go on to assume that the motion of the second particle, that is, its function $\mathbf{R}_2(t)$, is known, and we will write the generalized Jacobi equation (J_1) for the first particle when this particle is at the point $\mathbf{R} = \mathbf{R}_1(t)$. We will have

$$(J_1) \quad \frac{\partial \varphi_1}{\partial t} \equiv E_1 = \frac{1}{2m_1} (\mathrm{grad}_1\, \varphi_1)^2 + F_1 + F_{12} + Q_1 \tag{24}$$

with

$$Q_1 = \frac{\hbar^2}{2m_1} \left(\frac{\Box a_1}{a_1}\right)_{\mathbf{R}=\mathbf{R}_1} \simeq -\frac{\hbar^2}{2m_1} \left(\frac{\Delta a_1}{a_1}\right)_{\mathbf{R}=\mathbf{R}_1}. \tag{25}$$

In these equations, $\varphi_1(\mathbf{R}_1, \mathbf{r}_2, t)$ is the common phase of the u_1 and Ψ_1 waves of the first particle in the field of force created by the second —taking into account, if need be, the presence of obstacles causing interference and diffraction. $a_1(\mathbf{R}, \mathbf{r}_2, t)$ is the amplitude of the continuous Ψ_1 wave. The symbol $\mathrm{grad}_1\, \varphi_1$ signifies $(\mathrm{grad}\, \varphi_1)_{\mathbf{R}=\mathbf{R}_1}$, and $F_1(\mathbf{R}_1, t)$ represents an external potential that may possibly act upon the first particle, whereas $F_{12}(r_{12})$ represents the action of the second particle on the first.

In a like manner, let us now assume the motion of the first particle, that is, the function $\mathbf{R}_1(t)$, to be known, and let us write the generalized Jacobi's equation (J_2) for the second particle when it is at the point $\mathbf{R} = \mathbf{R}_2(t)$,

$$(J_2) \quad \frac{\partial \varphi_2}{\partial t} \equiv E_2 = \frac{1}{m_2} (\mathrm{grad}_2\, \varphi_2)^2 + F_2 + F_{21} + Q \tag{26}$$

with

$$Q_2 = \frac{\hbar^2}{2m_2} \left(\frac{\Box a_2}{a_2}\right)_{\mathbf{R}=\mathbf{R}_2} \simeq -\frac{\hbar^2}{2m_2} \left(\frac{\Delta a_2}{a_2}\right)_{\mathbf{R}=\mathbf{R}_2}. \tag{27}$$

In these equations, $\varphi_2(\mathbf{R}, \mathbf{r}_1, t)$ is the phase of the u_2 and Ψ_2 waves of the second particle in the field of force created by the first—making allowance, if need be, for the presence of obstacles causing interference and diffraction. The function $a_2(\mathbf{R}, \mathbf{r}_1, t)$ is the amplitude of the continuous Ψ_2 wave. The symbol $\text{grad}_2\, \varphi_2$ signifies $(\text{grad}\, \varphi_2)_{\mathbf{R}=\mathbf{R}_2}$. Finally $F_2(\mathbf{R}_2, t)$ represents an external potential that may possibly act on the second particle, while $F_{21}(r_{12})$ is the potential representing the action of the first particle on the second. We will assume, as usual, that $F_{12} = F_{21}$. As for the energy E_2 of the second particle, it is not, in general, constant.

Let us now view the matter from the point of view of configuration space where $\mathbf{R}_1$ with components x_1, y_1, z_1 and $\mathbf{R}_2$ with components x_2, y_2, z_2 become independent variables, and let us assume that the continuous wave $\Psi(\mathbf{R}_1, \mathbf{R}_2, t) = a(\mathbf{R}_1, \mathbf{R}_2, t)e^{\frac{i}{\hbar}\varphi(\mathbf{R}_1, \mathbf{R}_2, t)}$ obeys the Schrödinger equation.

The generalized Jacobi equation for the position $\mathbf{R}_1$, $\mathbf{R}_2$ of the representative point will be written

$$(\text{J}) \quad \frac{\partial \varphi}{\partial t} = E = \frac{1}{2m_1}(\text{grad}_1\, \varphi)^2 + \frac{1}{2m_2}(\text{grad}_2\, \varphi)^2 + F_1 + F_2 + F_{12} + Q \quad (28)$$

with

$$Q = \frac{\hbar^2}{2}\left(\frac{1}{m_1}\frac{\Box_1 a}{a} + \frac{1}{m_2}\frac{\Box_2 a}{a}\right)_{\mathbf{R}_1, \mathbf{R}_2} \simeq -\frac{\hbar^2}{2}\left(\frac{1}{m_1}\frac{\Delta_1 a}{a} + \frac{1}{m_2}\frac{\Delta_2 a}{a}\right)_{\mathbf{R}_1, \mathbf{R}_2}. \quad (29)$$

Here $\varphi(\mathbf{R}_1, \mathbf{R}_2, t)$ is the phase of the Ψ wave in configuration space, $a(\mathbf{R}_1, \mathbf{R}_2, t)$ is its amplitude. F_1, F_2 and F_{12} have the same significance as above.

Now, if we want the representation in three-dimensional physical space—the only physically real space—and the fictitious representation in configuration space to correspond, we must have the guidance formulas

$$\begin{aligned} m_1 \mathbf{v}_1 &= -\,\text{grad}_1\, \varphi_1 = -\,\text{grad}_1\, \varphi \\ m_2 \mathbf{v}_2 &= -\,\text{grad}_2\, \varphi_2 = -\,\text{grad}_2\, \varphi. \end{aligned} \quad (30)$$

But according to the lemma, this requires the following form for the phases:

$$\left.\begin{aligned}\varphi_1(\mathbf{R}_1, \mathbf{r}_{12}, t) &= \varphi_{11}(\mathbf{R}_1, t) + \varphi_{12}(\mathbf{r}_{12}, t)\\ \varphi_2(\mathbf{R}_2, \mathbf{r}_{12}, t) &= \varphi_{22}(\mathbf{R}_2, t) + \varphi_{21}(\mathbf{r}_{12}, t)\end{aligned}\right\}\varphi_{12} = \varphi_{21}$$
$$\varphi(\mathbf{R}_1, \mathbf{R}_2, \mathbf{r}_{12}, t) = \varphi_{11}(\mathbf{R}_1, t) + \varphi_{22}(\mathbf{R}_2, t) + \varphi_{12}(\mathbf{r}_{12}, t). \tag{31}$$

In this way we obtain a precise statement of the general form, valid even in the cases of external actions or of obstacles to propagation, of the phases φ_1 and φ_2 of the individual waves (u and Ψ) associated with each of the two particles, as well as the phase φ of the fictitious Ψ wave in configuration space.

It may be pointed out that there does not seem any necessary reason why the function $\varphi_{12} = \varphi_{21}$ should be solely a function of the distance

$$r_{12} = \sqrt{(x_1 - x_2)^2 + (y_1 - y_2)^2 + (z_1 - z_2)^2};$$

it would suffice if it were the function of the three components $x_1 - x_2$, $y_1 - y_2$, $z_1 - z_2$ of the vector $\mathbf{r}_{12}$.

Moreover, the "quantum forces" must have the same values, whether they are calculated in physical space or in configuration space, and this fact requires the conditions

$$\text{grad}_1\, Q_1 = \text{grad}_1\, Q; \qquad \text{grad}_2\, Q_2 = \text{grad}_2\, Q \tag{32}$$

and by applying the lemma we obtain further

$$\left.\begin{aligned}Q_1(\mathbf{R}_1, \mathbf{r}_{12}, t) &= Q_{11}(\mathbf{R}_1, t) + Q_{12}(\mathbf{r}_{12}, t)\\ Q_2(\mathbf{R}_2, \mathbf{r}_{12}, t) &= Q_{22}(\mathbf{R}_2, t) + Q_{21}(\mathbf{r}_{12}, t)\end{aligned}\right\}Q_{12} = Q_{21}$$
$$Q(\mathbf{R}_1, \mathbf{R}_2, t) = Q_{11}(\mathbf{R}_1, t) + Q_{22}(\mathbf{R}_2, t) + Q_{12}(\mathbf{r}_{12}, t). \tag{33}$$

We can make the same remark about $Q_{12} = Q_{21}$ that we made for φ_{12}: There does not seem any necessary reason why that function must depend on the distance $\mathbf{r}_{12}$, it is sufficient for it to be a function of the three components $x_1 - x_2$, $y_1 - y_2$, $z_1 - z_2$ of the vector $\mathbf{r}_{12}$. This observation makes it possible to modify my 1927 hypothesis about Q_{12}.

Formulas (31) and (33) show that the transition from φ_1 and φ_2 to φ, and from Q_1 and Q_2 to Q must be effected in exactly the same way as the transition from $\mathscr{L}_1$ and $\mathscr{L}_2$ to $\mathscr{L}$, and from E_1 and E_2 to E in the Classical Mechanics of systems.

So now, by comparing formulas (24), (26) and (28) and by taking (31) into account, we obtain

$$E = E_1 + E_2 - F_{12} + Q - Q_1 - Q_2 \tag{34}$$

or, according to (33)

$$\begin{aligned} E &= E_1 + E_2 - F_{12} - Q_{12} \\ &= \tfrac{1}{2}m_1 v_1^2 + \tfrac{1}{2}mv_2^2 + F_1 + F_2 + F_{12} + Q_1 + Q_2 + Q_{12}. \end{aligned} \tag{35}$$

This formula, moreover, appears quite natural because it treats the ordinary interaction potential F_{12} and the "quantum interaction potential" Q_{12} in a symmetrical manner.

The method we have just outlined has the advantage of giving us a better analysis of the general form that the phases and quantum potentials must have. But it still does not supply a complete justification of the passage from the Wave Mechanics of a single particle in a given field to the Wave Mechanics of systems of particles in interaction within the framework of the theory of the Double Solution. To obtain equations (30) and (32), we have *assumed* that there is an agreement between the representation of the motion by two u waves in physical space and the representation of the motion by the fictitious Ψ wave in configuration space. Now it is precisely this agreement that must be rigorously justified. So we are going to indicate a kind of reasoning that seems to bring us much closer to our objective.

4. A comparison of the relative motion of two interacting particles with the representation of that system's motion in configuration space

We now consider once again, and exclusively, an isolated system formed of two particles 1 and 2 whose interaction is represented by a potential $V(r)$, which is a function of the distance r only. In order to study it, we can still adopt a Galilean reference system in which the center of gravity is immobile and used, for example, as the origin of the coördinates. Passage into another Galilean system will then merely introduce in addition the rectilinear and uniform motion of the center of gravity. We will first study the case where the Newtonian approximation of Classical Mechanics (that is, from the wave-mechanical viewpoint, the approximation of Geometrical Optics) is valid. We will then try to proceed to the general case.

a. The approximation of Geometrical Optics

Whenever Classical Mechanics is valid, one can write Newton's equation of motion for the two particles

$$m_1 \mathbf{a}_1 = -\operatorname{grad}_1 V; \qquad m_2 \mathbf{a}_2 = -\operatorname{grad}_2 V. \tag{36}$$

The two then describe in physical space the trajectories L_1 and L_2 with well defined motions. The two trajectories and the corresponding motions are "correlated" in a one-to-one way.

But one can also represent the combined motion of the two particles by the displacement of a representative point in configuration space $x_1, \ldots, z_2$ and write Jacobi's equation in the space

$$\frac{\partial S}{\partial t} = \frac{1}{2m_1} (\text{grad}_1 S)^2 + \frac{1}{2m_2} (\text{grad}_2 S)^2 + V(r) \tag{37}$$

where $\partial S/\partial t$ has a constant value E, the total energy of the system. To every complete integral $S(x_1, \ldots, z_2, t, \alpha, \beta, \ldots)$ of equation (37) corresponds a trajectory L of the representative point which is *one* of the curves orthogonal to the surfaces S = const. in configuration space, and the motion of the particles along the corresponding trajectories L_1 and L_2 in physical space is given by Jacobi's formulas

$$\mathbf{v}_1 = -\frac{1}{m_1} \text{grad}_1 S, \qquad \mathbf{v}_2 = -\frac{1}{m_2} \text{grad}_2 S. \tag{38}$$

Thus the representation in configuration space makes the ensemble of those motions of the particle belonging to the same class corresponding to the complete integral S.

In order to study the combined motions of two particles one can also use another method in Classical Mechanics which is entirely different from the preceding methods and which consists in considering the relative motion of one of the particles with respect to the other. It is well known that, if we take a system of axes with fixed directions having one of the particles as its origin, the motion of the other particle in this *non-Galilean* reference system will be the same as if the system were Galilean and the other particle possessed a "reduced mass" such that

$$\mu = \frac{m_1 m_2}{m_1 + m_2} \cdot \tag{39}$$

In other words, the influence of the non-uniform motion of the new reference system is indicated solely by an apparent variation in the mass of the particle in relative motion. In the non-Galilean system fixed to one of the particles, one can then write for the motion of the other particle the equation of the Newtonian type

$$\mu \mathbf{a}^* = - \text{grad}^* V \tag{40}$$

where the asterisk indicates that a quantity is evaluated in the non-Galilean system.

To be more specific, let us suppose that we take particle 2 as the origin of the non-Galilean reference system. We see immediately that we will have as Jacobi's equation of particle 1 in this system

$$\frac{\partial S_1^*}{\partial t} = \frac{1}{2\mu} (\text{grad}^* \, S_1)^2 + V(r). \tag{41}$$

The relative motion of particle 1 corresponding to a complete integral of (41) will take place along a trajectory L_1^* which will be a curve orthogonal to the surfaces $S_1^* =$ const. and the motion along L_1^* will be given by the equation

$$\mathbf{v}_1^* = -\frac{1}{\mu} \text{grad}^* \, S_1^*. \tag{42}$$

We must now examine very closely the meaning of these equations which are valid in the relative system.

Whereas in the Galilean system bound to the center of gravity the partial energies $E_1 = T_1 + V$ and $E_2 = T_2 + V$ of the two particles in motion were variable, and whereas only the total energy $E = T_1 + T_2 + V$ was constant, the energy $E_1^* = \partial S_1^*/\partial t$ in the relative motion is constant, as one could deduce by the usual manner starting from equation (40). This derives from the fact that in passing into the system fixed to 2, we have concentrated all the energy of the combination of the two particles on the particle that remains in motion. One has, moreover, $E_1^* = E$, that is

$$\tfrac{1}{2}\mu(\mathbf{v}_1 - \mathbf{v}_2)^2 + V = \tfrac{1}{2}m_1 v_1^2 + \tfrac{1}{2}m_2 v_2^2 + V \tag{43}$$

as is easy to verify.

Naturally, if we had taken particle 1 as the origin for the relative coördinates by changing the variables to $x^* = x - x_1(t) \ldots$, we would have found for particle 2 in this sytem Jacobi's equation

$$\frac{\partial S_2^*}{\partial t} = \frac{1}{2\mu} (\text{grad}^* \, S_2^*)^2 + V(r) \tag{44}$$

and the energy $E_2^* = \partial S_2^*/\partial t$ would also have been a constant equal to E.

The two Jacobi functions S_1^* and S_2^*, obeying the two equations (41) and (44) of the same form, will be equal to the same function $F(\mathbf{r}^*, t)$ of the relative coordinates and of time.

We come now to the all-important point. Let us consider a certain complete integral of equation (41), $S_1^*(\mathbf{r}^*, t)$. The various curves orthogonal to the surfaces $S_1^* =$ const. and the corresponding motions defined by (42) represent possible motions of particle 1 around particle 2. Let L_1^* be one of the possible trajectories; by returning to the Galilean system where the center of gravity is at the origin of the coördinates, and by using the relation $m_1\mathbf{r}_1 + m_2\mathbf{r}_2 = 0$, one can deduce from the motion L_1^* the correlated motions L_1 and L_2 of the two particles around the center of gravity. Among the curves (C) forming the congruence of the normals to the surface $S_1^* =$ const. in the non-Galilean reference system fixed to 2, only one of these curves is in reality described by particle 1 in its relative motion. However, since in the non-Galilean system fixed to 2, particle 2 plays only the role of a simple center of forces, *we must consider that equation* (41) *gives us, in the approximation of Geometrical Optics, the propagation of the phase* S_1^* *of the wave* u_1^* *of particle* 1 *throughout this entire reference system.*

One thus sees that, in the relative reference system function S_1^* gives us *simultaneously* the ensemble of possible motions of the same class L_1^* of particle 1 in the central static field created by particle 2 and the phase throughout the whole relative system of the u_1^* wave of particle 1 when it describes *any* of the trajectories L_1^*. This basic obervation is the key to the argument here presented.

Naturally, if we assign the motion of particle 2 to particle 1, it would be the function $S_2^*(\mathbf{r}^*, t) = S_1^*(\mathbf{r}^*, t)$ which would serve to represent the totality of possible motions of particle 2 in the central static field created by particle 1 and also, throughout the whole relative system, the phase of the u_2^* wave of particle 2 when it describes any one of the trajectories.

Let us now consider the Jacobi function S for the two particles in configuration space. It is a function $S(\mathbf{R}_1, \mathbf{R}_2, t)$ of the six configuration variables $x_1, \ldots, z_2$ and of time t that obeys the Jacobi equation (37). If we introduce the relative variables $x^* = x_1 - x_2, \ldots$, we will have

$$\frac{\partial S}{\partial x_1} = \frac{\partial S}{\partial x^*}, \quad \frac{\partial S}{\partial x_2} = -\frac{\partial S}{\partial x^*}$$

the variables $x_1, \ldots, z_2$ being connected, moreover, by the relations $m_1x_1 + m_2x_2 = 0, \ldots$ which express the stationary position of the center of gravity. The Jacobi equation for S then takes on the form, in the relative system,

$$\frac{\partial S^*}{\partial t} = \frac{1}{2\mu}[(\text{grad}^* S^*)^2] + V(r). \tag{45}$$

The identity of the form of equations (41), (44) and (45) thus shows that the Jacobi function S of the two-particle system (which is equal to the phase of the Ψ wave in configuration space in the approximation of Geometrical Optics) has in the system fixed to particle 2 the same expression $S^* = F(\mathbf{r}, t)$ as has S_1, and in the system fixed to particle 1 the same expression $F(\mathbf{r}, t)$ as has S_2.

Coming back to the Galilean system fixed to the center of gravity, it is seen that by using the notation of the preceding section one has

$$S_1(\mathbf{R}_1, \mathbf{r}_{12}, t) = S_2(\mathbf{R}_2, \mathbf{r}_{12}, t) = S(\mathbf{R}_1, \mathbf{R}_2, t) = F(r_{12}, t) \tag{46}$$

and at the correlated points of L_1 and L_2, the formulas

$$m_1\mathbf{v}_1 = -\text{grad}_1 S_1 = -\text{grad}_1 S, \quad m_2\mathbf{v}_2 = -\text{grad}_2 S_2 = -\text{grad}_2 S \tag{47}$$

are clearly valid. Formulas (46) clearly correspond to formulas (31) of the preceding section with $\varphi_{11} = \varphi_{22} = 0$ and $\varphi_{12} = S_{12} = F$ (the zero value of φ_{11} and of φ_{22} resulting from the immobility of the center of gravity in the Galilean system). If we adopted a Galilean system of reference in which the center of gravity would have a uniform rectilinear motion of energy E_g and momentum $\mathbf{P}$ it would be necessary to add the term $E_g t - \mathbf{P}\cdot\mathbf{R}$ where $\mathbf{R} = (m_1\mathbf{r}_1 + m_2\mathbf{r}_2)/(m_1 + m_2)$ etc. to S. Formulas (46) would then take on the form

$$\left.\begin{aligned} S_1(\mathbf{R}_1, \mathbf{r}_{12}, t) &= S_{11}(\mathbf{R}_1, t) + S_{12}(\mathbf{r}_{12}, t) \\ S_2(\mathbf{R}_2, \mathbf{r}_{12}, t) &= S_{22}(\mathbf{R}_2, t) + S_{21}(\mathbf{r}_{12}, t) \end{aligned}\right\}(S_{12} = S_{21}) \\ S(\mathbf{R}_1, \mathbf{R}_2, t) = S_{11}(\mathbf{R}_1, t) + S_{22}(\mathbf{R}_2, t) + S_{12}(\mathbf{r}_{12}, t) \tag{48}$$

which coincides exactly, in the approximation of Geometrical Optics, with formulas (31) of the preceding section.

Let us add one interesting point: Equation (41), (44) and (45) may be written

$$E_1^* = T_1^* + V \qquad E_2^* = T_2^* + V \qquad E = T^* + V \tag{49}$$

with

$$E = E_1^* = E_2^* \qquad T^* = T_1^* = T_2^* = T_1 + T_2. \tag{50}$$

Whence

$$E_1^* = (T_1 + V) + T_2 = E_1 + T_2, \quad E_2^* = (T_2 + V) + T_1 = E_2 + T_1, \\ E_1 + E_2 = E_1^* + E_2^* - (T_1 + T_2) = 2E - (E - V) \tag{51}$$

or

$$E = E_1 + E_2 - V. \tag{52}$$

Thus we again have the expression previously pointed out for the total energy E of the system of two particles as a function of their individual energies E_1 and E_2, and we see more clearly the reason why we do not have $E = E_1 + E_2$.

We have been reasoning exclusively within the approximation of Geometrical Optics—a procedure that had the advantage of allowing us to remain on the solid ground of Classical Mechanics. Now, by similar arguments, we will seek to extrapolate the results obtained into regions beyond the limits of Geometrical Optics by introduction of the ideas of the causal theory.

b. A study of the same problem outside the limits of Geometrical Optics

Let us first write the equation of propagation of the Ψ wave in configuration space

$$\frac{\hbar}{i}\frac{\partial \Psi}{\partial t} = -\frac{\hbar^2}{2m_1}\Delta_1 \Psi - \frac{\hbar^2}{2m_2}\Delta_2 \Psi + V(r)\Psi. \tag{53}$$

If we express Ψ in the form

$$\Psi = a(x_1 \ldots z_2, t)e^{\frac{i}{\hbar}\varphi(x_1,\ldots,z_2,t)}$$

the quantity $E = \partial\varphi/\partial t$ will have to be the constant energy of the system. In the double-solution view, the particles in the system must be well localized and must describe in physical space correlated trajectories L_1 and L_2 represented in configuration space by the trajectory L of the representative point—a trajectory that is *one* of the curves orthogonal to the surfaces $\varphi = \text{const}$. The motions of the particles along trajectories L_1 and L_2 in physical space would be given by the guidance formulas

$$\mathbf{v}_1 = -\frac{1}{m_1}\text{grad}_1\,\varphi, \qquad \mathbf{v}_2 = -\frac{1}{m_2}\text{grad}_2\,\varphi. \tag{54}$$

In this way the representation in configuration space would always make the ensemble of correlated motions of the same class correspond to the phase φ.

As previously, it is natural to consider the relative motions of each of the particles with respect to the other one. It is easy to demonstrate that, in the non-Galilean system fixed to one of the two particles, one has for the Ψ wave the equation of propagation

$$\frac{\hbar}{i}\frac{\partial \Psi^*}{\partial t} = -\frac{\hbar^2}{2\mu}\Delta^* \Psi^* + V(r)\Psi^* \tag{55}$$

μ always having the value (39), which, in the approximation of Geometrical Optics, would bring us back to the equation of Jacobi (45) with $\varphi^* = S^*$.

What should be, from the point of view of the Double Solution, the equation of propagation (outside the singular region) of the u_1 wave of particle 1 in the reference system fixed to particle 2? If we write it in the form

$$u_1^*(x^*, y^*, z^*, t) = a_1^*(x_1^*, y_1^*, z_1^*, t)e^{\frac{i}{\hbar}\varphi_1^*(x_1^*, y_1^*, z_1^*, t)}$$

we know that within the approximation of Geometrical Optics where $\varphi_1^* = S_1^*$ we must once more end up with equation (41). This leads us to write as the equation of propagation of u_1^*, in the relative non-Galilean system

$$\frac{\hbar}{i}\frac{\partial u_1^*}{\partial t} = -\frac{\hbar^2}{2\mu}\Delta^* u_1^* + V(r)u_1^*. \tag{56}$$

The relative motion will take place along one of the curves orthogonal to the surfaces $\varphi_1 = \text{const.}$ with velocity

$$\mathbf{v}_1 = -\frac{1}{\mu}\,\text{grad}^*\,\varphi_1^*.$$

Here again, for the reasons set forth above, $E_1^* = \partial\varphi_1^*/\partial t$ will be equal to the constant energy E of the system, the entire energy of the system being concentrated on the relative motion of particle 1.

Naturally, had we taken particle 1 as the origin of the coördinates of the relative system, we would have been obliged to take for the equation of propagation of the u_1^* of particle 2 in this system

$$\frac{\hbar}{i}\frac{\partial u_2^*}{\partial t} = -\frac{\hbar^2}{2\mu}\Delta^* u_2^* + V(r)u_2^* \tag{57}$$

which, within the approximation of Geometrical Optics, would again give us equation 44) with $S_2^* = \varphi_2^*$. Since the functions u_1^* and u_2^* obey equations of the same form, they may be regarded as equal to the same function of $\mathbf{r}^*$ and t, and this leads us to give them the common expression $a^*(\mathbf{r}^*, t)e^{\frac{i}{\hbar}\varphi^*(\mathbf{r}^*,t)}$.

In the non-Galilean reference system fixed to particle 2, this particle no longer plays the role of a center of force, and we are brought back to the case of the motion of a particle in a given field, with the wave $u_1^* = a_1^* e^{\frac{i}{\hbar}\varphi_1^*}$ obeying (56). The ensemble of curves orthogonal to the surfaces $\varphi_1 =$ const., and the motions defined by $\mathbf{v}_1^* = -(1/\mu)\,\text{grad}^*\varphi_1^*$ represent in this system the ensemble of possible motions of the same class. If L_1^* is the trajectory described by particle 1, one will be able, by returning to the Galilean system fixed to the center of gravity and by utilizing the relation $m_1\mathbf{r}_1 + m_2\mathbf{r}_2 = 0$, deduce from the motion L_1^* the correlated motions L_1 and L_2 of the two particles around the center of gravity.

We see that, in the relative system, the function $a^* e^{\frac{i}{\hbar}\varphi^*}$ represents *both* the ensemble of possible L_1^* motions of the same class for particle 1 and (except in the singular region) the wave of particle 1 when it describes any one of the trajectories L_1^*. That is the all-important point.

Naturally, if we assign the motion of particle 2 to particle 1, it will be the function $a^* e^{\frac{i}{\hbar}\varphi^*}$ that will serve to represent both the ensemble of trajectories of the same class L_2^* and the u_2^* wave associated with particle 2 when it describes any *one* of the trajectories L_2^*.

If we now compare equations (56) and (57), equation (55) being satisfied by the Ψ wave in configuration space, we see that for the Ψ^* function the same expression $a^*(\mathbf{r}^*, t)e^{\frac{i}{\hbar}\varphi^*(\mathbf{r}^*, t)}$ may be chosen as we did for the functions u_1^* and u_2^*.[2] We conclude from this that formulas (31) and (33) of the preceding paragraph, where φ_{11} and φ_{22} are zero as well as Q_{11} and Q_{22}, and which are reduced to

$$\begin{aligned} \varphi_1(\mathbf{r}_{12}, t) &= \varphi_2(\mathbf{r}_{12}, t) = \varphi(\mathbf{r}_{12}, t) = \varphi_{12}(\mathbf{r}_{12}, t) \\ Q_1(\mathbf{r}_{12}, t) &= Q_2(\mathbf{r}_{12}, t) = Q(\mathbf{r}_{12}, t) = Q_{12}(\mathbf{r}_{12}, t) \end{aligned} \tag{58}$$

are verified.

The zero value of φ_{11} and φ_{22} results from the immobility of the center of gravity of the chosen Galilean system. In some other Galilean system in which the center of gravity would move in a straight line with con-

[2] The result obtained may be expressed by saying that, in the relative reference system, where one of the particles plays only the role of a center of forces, the regular part of the moving particle's u wave coincides (to within a constant of normalization) with its Ψ wave. This result will appear again in Chapter XVII.

stant energy E_g and momentum $\mathbf{p}$, the term $E_g t - \mathbf{P} \cdot \mathbf{R}$ would be introduced into the phase and we would once more be back at formulas (31).

Let us further add that the verification of the expression $E = E_1 + E_2 - V - Q_{12}$ is obtained here in the same way that we obtained the expression for the formula $E = E_1 + E_2 + V$ above. The generalized Jacobi equations corresponding to the equations of propagation (55), (56) and (57) give us, on substituting $Q = Q_{12}$,

$$E_1^* = T_1^* + V + Q; \quad E_2^* = T_2^* + V + Q; \quad E = T^* + V + Q; \qquad (59)$$

with

$$E = E_1^* = E_2^* \qquad T^* = T_1^* = T_2^* = T_1 + T_2. \qquad (60)$$

We deduce from this

$$E_1^* = (T_1 + V + Q) + T_2 = E_1 + T_2; \quad E_2^* = (T_2 + V + Q) + T_1 = E_2 + T_1;$$
$$E_1 + E_2 = E_1^* + E_2^* - (T_1 + T_2) = 2E - (E - V - Q) \qquad (61)$$

whence

$$E = E_1 + E_2 - V - Q. \qquad (62)$$

The foregoing reasoning is valid for two particles in the absence of external fields.

If there are external fields, it is still easy to obtain formulas (31) and (33) in the case where—providing these fields are approximately constant throughout the extent of the system—there is a separation of the motion of the center of gravity from the relative motion. If this condition is not fulfilled, the problem is more complicated and would require re-examination. The same is true for the case of more than two particles, where it would perhaps be necessary to employ methods similar to those used in Mathematical Astronomy in the n-body problem. However that may be, the method employed in this section is very instructive and seems well adapted to the solution of the problem under consideration.

5. The case of particles of the same nature

The case of systems containing particles of the same physical nature poses an even more difficult problem. In particular, one may wonder whether retaining the notion of a trajectory postulated by the theory of the Double Solution is compatible with the indistinguishability of particles assumed by present-day quantum theory. It does seem that the experimental verification of Bose-Einstein statistics for bosons furnishes direct proof of the indistinguishability of these particles, and the question is to decide whether this experimental result is compatible with the conceptions of the Double Solution. Relying on the considerations developed above, we will indicate how the question may be approached.

Let us continue to consider a system of two particles. We will assume them to be of the same nature, so that $m_1 = m_2$. The study of such a system in standard Wave Mechanics shows that, if the regions of possible presence of the two particles overlap, it is necessary to assume that the Ψ wave of the system in configuration space is either symmetrical or anti-symmetrical.[3]

Since the wave equations of the two particles are the same, it is natural to assume that, if the u wave-trains partially overlap, the waves may be superposed and form a single u wave that may be represented by the formula

$$u(x, y, z, t) = f(x, y, z, t)e^{\frac{i}{\hbar}\varphi(x, y, z, t)} \tag{63}$$

the amplitude $f(x, y, z, t)$ here having two distinct mobile singular regions. In other words, the respective u waves of the two particles $u_1(\mathbf{R}, \mathbf{r}_{12}, t)$ and $u_2(\mathbf{R}, \mathbf{r}_{12}, t)$ would fuse into a single wave with two singular regions. It would be this sort of fusion of u waves, possible for bosons, that would explain the statistics of these particles.

Let us assume that one of the particles is situated at the point $\mathbf{R}_1$, with the other having a position determined by $\mathbf{r}_{12}$. Since there must be a unique phase of the u wave, we must have

$$\varphi_1(\mathbf{R}_i, \mathbf{r}_{12}, t) = \varphi_2(\mathbf{R}_i, \mathbf{r}_{12}, t) \qquad (i = 1, 2) \tag{64}$$

whence, according to formulas (31)

$$\varphi_{11}(\mathbf{R}_i, t) = \varphi_{22}(\mathbf{R}_i, t) \tag{65}$$

and, further,

$$\varphi(\mathbf{R}_1, \mathbf{R}_2, \mathbf{r}_{12}, t) = \varphi_{11}(\mathbf{R}_1, t) + \varphi_{22}(\mathbf{R}_2, t) + \varphi_{12}(\mathbf{r}_{12}, t), \tag{66}$$

$\varphi(\mathbf{R}_1, \mathbf{R}_2, \mathbf{r}_{12}, t)$ still being the phase of the Ψ wave in configuration space.

Similarly, for the amplitudes of the respective Ψ waves, we will have to have

$$a_1(\mathbf{R}_1, \mathbf{r}_{12}, t) = a_2(\mathbf{R}_2, \mathbf{r}_{12}, t) \tag{67}$$

[3] See Chapter IV, section 4.

and in this way one obtains from the definition of the quantum potentials and from formulas (33)

$$\mathcal{Q}_{11}(\mathbf{R}_i, t) = \mathcal{Q}_{22}(\mathbf{R}_i, t) \tag{68}$$

$$\mathcal{Q}(\mathbf{R}_1, \mathbf{R}_2, \mathbf{r}_{12}, t) = \mathcal{Q}_{11}(\mathbf{R}_1, t) + \mathcal{Q}_{22}(\mathbf{R}_2, t) + \mathcal{Q}_{12}(\mathbf{r}_{12}, t).$$

These formulas, like (65) and (66), result, moreover, from the fact that the singular regions are indistinguishable and may be permuted without any resultant modification of the wave phenomenon represented by u.

So the quantum potential of configuration space

$$\mathcal{Q} = \frac{\hbar^2}{2m_1} \frac{[\Box_1 + \Box_2]a(\mathbf{R}_1, \mathbf{R}_2, \mathbf{r}_{12}, t)}{a(\mathbf{R}_1, \mathbf{R}_2, \mathbf{r}_{12}, t)} \simeq -\frac{\hbar^2}{2m_1} \frac{[\Delta_1 + \Delta_2]a}{a} \tag{69}$$

must be symmetrical in $\mathbf{R}_1$ and $\mathbf{R}_2$.

If $\mathscr{A}(\mathbf{R}_1, \mathbf{R}_2, \mathbf{r}_{12}, t)$ designates the amplitude of any solution of the equation of the waves in configuration space, $\mathscr{A}(\mathbf{R}_2, \mathbf{R}_1, \mathbf{r}_{12}, t)$ will likewise be a solution, and one will be obliged to form a linear combination of the form

$$\mathrm{C}\mathscr{A}(\mathbf{R}_1, \mathbf{R}_2, \mathbf{r}_{12}, t) + \mathrm{D}\mathscr{A}(\mathbf{R}_2, \mathbf{R}_1, \mathbf{r}_{12}, t) \tag{70}$$

such that the quantity

$$\frac{\mathrm{C}\Box\mathscr{A}(\mathbf{R}_1, \mathbf{R}_2, \mathbf{r}_{12}, t) + \mathrm{D}\Box\mathscr{A}(\mathbf{R}_2, \mathbf{R}_1, \mathbf{r}_{12}, t)}{\mathrm{C}\mathscr{A}(\mathbf{R}_1, \mathbf{R}_2, \mathbf{r}_{12}, t) + \mathrm{D}\mathscr{A}(\mathbf{R}_2, \mathbf{R}_1, \mathbf{r}_{12}, t)} \sim \mathcal{Q}. \tag{71}$$

where $\Box = \Box_1 + \Box_2$ is in no way modified by the permutation of $\mathbf{R}_1$ and of $\mathbf{R}_2$, that is, in the permutation of the position of two singular regions. On writing this condition, one easily sees that $\mathrm{C}^2 = \mathrm{D}^2$, that is, $|\mathrm{C}| = |\mathrm{D}|$ and $2 \arg \mathrm{C} = 2 \arg \mathrm{C} + 2\pi n$, whence

$$\mathrm{C} = |\mathrm{C}|e^{i\alpha} \qquad \mathrm{D} = \pm |\mathrm{C}|e^{i\alpha} = \pm \mathrm{C}.$$

One is thus led to allow, for the Ψ wave in configuration space, only symmetrical and antisymmetrical solutions, in conformity with the well known result of Wave Mechanics of similar particles. But here this result appears as a consequence of the fact that the u waves of the similar particles, when they overlap in space, fuse into a single wave involving several singular regions whose roles must be interchangeable; for, in a single u wave, these singular regions are identical, and their permutation can produce no effect. The picture thus obtained for the u wave strikes us as being of very great interest.

With the classical result thus achieved as our basis, it must be shown why the symmetrical solution fits bosons and the antisymmetrical solution fits fermions and, in the case of the fermions, we must justify the validity of Pauli's exclusion principle. So it becomes necessary to show why the singular "boson" regions can be grouped together in a cluster on one and the same u wave, while singular "fermion" regions cannot be treated that way and always remain isolated on their respective u waves. In order to understand this point clearly, it would certainly be necessary to introduce the notion of spin, which is essential to the rigorous definition of symmetrical and antisymmetrical wave functions. Thus, the theory of the Double Solution can approach this question only after it has been extended to the case of particles with spin other than zero, and especially to the case of the Wave Mechanics of the Dirac electron—which will not be done until Chapter XVI. But, even after such an extension, we will not be able to solve the problem that is here posed, for it is one of those problems which we must, at the present moment, reserve for future research.[4]

[4] The arguments set forth in this chapter are certainly inadequate and do not give a complete solution of this difficult problem. Since this book was written, I have again taken up this question in collaboration with J. L. Andrade e Silva, and we have obtained results that permit us to hope that the problem can be completely solved in a satisfactory manner.

On this subject, see the following notes in *Comptes rendus de l'Académie des Sciences:* 244 (1957) 529; 245 (1957) 1893 and 2018; 246 (1958) 391; 248 (1959) 1785, 1947 and 2291.

Chapter XIII

THE PROBABILISTIC SIGNIFICANCE OF $|\Psi|^2$ AND ITS JUSTIFICATION

1. A reconsideration of the 1927 arguments

In 1927, I had, as we have seen, tried to justify the attribution of the following significance to the quantity $a^2 = |\Psi|^2$: It represents the probability of presence of a singularity-particle at the instant t at the corresponding point in space. In order to effect this justification I started with the case of the plane monochromatic wave in the absence of a field (physically one should say: the case of a very extended wave-train identifiable throughout almost its entirety with a plane monochromatic wave). The possible trajectories are then straight lines parallel to the direction of propagation of the wave. We cannot know (unless we make an observation that would completely alter the situation) in which of these trajectories the particle lies, nor at what point it is situated along that trajectory. For this reason, I said, one seems justified in considering all the positions of the particle in the wave-train as equally probable; and this permits us to adopt the expression $a^2 = |\Psi|^2$, which is here constant for the probability of presence. If the wave-train then enters a region where any field of force whatever is present, the continuity equation

$$\frac{\partial a^2}{\partial t} + \operatorname{div}(a^2 \mathbf{v}) = 0$$

(where $\mathbf{v}$ is given by the guidance formula) will make it possible to conclude that the probability of presence must always remain equal to $a^2 = |\Psi|^2$. Naturally, this reasoning must also apply to the Ψ wave of a system in configuration space, interpreting the Ψ wave as we have just done in the preceding chapter. In this case one may begin with a state of the system where the constituent parts associated with trains of Ψ waves of finite dimensions are far enough apart from each other to be represented by a train of plane monochromatic waves in configuration space. It is then natural to admit still further that, at the outset, the probability of presence of the representative point is given by the constant $a^2 = |\Psi|^2$, and one concludes from this that, throughout the duration of the interaction, this probability of presence will remain equal to $|\Psi|^2$. A natural consequence of this is that, if the particles are once more separated and no longer interacting after the conclusion of the interaction, the probability of each of them in physical space will still be deducible from the value of $|\Psi|^2$.

Consider as an example an electron which, in its initial state, is in uniform rectilinear motion outside of any field. If it is then involved in interactions with other particles, with an attendant exchange of energy and momentum, and if it finally enters a stationary state inside an atom, the foregoing considerations seem to justify viewing its final probability of presence at various points within the atom as being given by the square of the modulus of the Ψ function representing its final stationary state.

Thus, if we admit that, beginning with an initial rectilinear and uniform state of motion for a particle, one can always—by means of the appropriate interactions—end up with the particle in any final state, one can justify the identification of $|\Psi|^2$ with the probability of presence, and one may do so for the general case. However, this conclusion rests on the following hypothesis: When the Ψ wave is made up of a wave-train that is approximately a plane monochromatic wave, it is legitimate to consider the probability of presence as given by $a^2 = |\Psi|^2$.

But this hypothesis may appear arbitrary. In a general way, the density of the probability of presence $\rho(x, y, z, t)$ must obey, as the a^2 function also must, the continuity equation

$$\frac{\partial \rho}{\partial t} = - \operatorname{div} \rho \mathbf{v} \tag{1}$$

where $\mathbf{v}(x, y, z, t)$ is a known function of x, y, z, t. This equation being of the first order in t, ρ is completely determined if we know the initial form $\rho(x, y, z, t_0)$ at the original instant t_0. If we put

$$\rho(x, y, z, t_0) = a^2(x, y, z, t_0) = |\Psi(x, y, z, t_0)|^2 \tag{2}$$

then the solution of (1) is $\rho = a^2 = |\Psi|^2$. But choice (2) of the initial form is arbitrary, and one could make an infinite number of other choices that would be just as acceptable. Even in the case of the plane monochromatic wave, the choice of $\rho = a^2 = |\Psi|^2$, even though quite natural, has an arbitrary character.

Nevertheless, in order to defend the validity of this hypothesis which formed the basis of my 1927 arguments, one may point out that, whenever one wishes to apply the calculus of probabilities to a concrete problem, there is always an arbitrary hypothesis of a physical nature forming the very basis of that application. Even in the extremely simple and intuitive case of heads-or-tails, when we assign the same $\frac{1}{2}$ probability to the "head" and to the "tail" hypothesis respectively, we implicitly assume the perfect symmetry of the coin. If we consider a homogeneous cloud of droplets falling normally on a plane surface, we naturally, assume that equal areas of the plane enjoy the same probability of receiving a droplet, that is, we assume that the number of vertical trajectories of the drops striking a surface $d\sigma$ of the plane is proportional to $d\sigma$. That is exactly the same sort of hypothesis we made in the case of the plane monochromatic wave, for, if we consider the ensemble of the particle's possible parallel trajectories that strike a wave-surface, the hypothesis made by putting $\rho = a^2 =$ const. would amount precisely to admitting that the number of possible trajectories traversing a small area $d\sigma$ of the wave surface is proportional to $d\sigma$. The necessity of adopting, if statistical considerations are to be introduced, a basic postulate that is unavoidably somewhat arbitrary seems to justify the very natural hypothesis we have made. My argument, therefore, seems to me to retain considerable force.

Nevertheless, since the force of the argument may be contested, it is interesting to examine the problem from other points of view.

2. A comparison with Liouville's theorem and the ergodic theory

Classical Statistical Mechanics rests on Liouville's theorem, which is derived from Hamilton's equations. Let us first point out that since Hamilton's equations are valid for the motion of particles in the causal theory (as we have shown), Liouville's theorem is equally valid and may also be utilized in the problem before us.

Let us recall Liouville's theorem: Let there be a collection of N particles with coördinates q_i; having conjugate momenta p_i. Such a system may be represented by a point in a space having 6N dimensions constituted by the q_i and p_i. A collection of systems of the same sort will then be represented by a "cloud" of representative points in this space, to which we give the name, since the time of Gibbs, extension-in-phase. The cloud of representative points is similar to a fluid in motion in extension-in-phase. One can define its density $\sigma(q_i, p_i, t)$ and its velocity **V** at every point, **V** being a vector of 6N components given by the ensemble of $\dot{q}_i = dq_i/dt$ and $\dot{p}_i = dp_i/dt$.

Since the number of systems under consideration remains unchanged, the fluid in question must obey the continuity equation

$$\frac{\partial \sigma}{\partial t} + \text{div}\,(\sigma \mathbf{V}) = \frac{\partial \sigma}{\partial t} + \sigma\,\text{div}\,\mathbf{V} + \mathbf{V} \cdot \text{grad}\,\sigma = 0 \tag{3}$$

—the operators div and grad here being defined in a 6N-dimensional extension-in-phase. Now, by assuming that for every index i the following Hamiltonian equations are valid

$$\dot{q}_i = \frac{\partial \mathrm{H}}{\partial p_i}, \qquad \dot{p}_i = -\frac{\partial \mathrm{H}}{\partial q_i} \tag{4}$$

one sees that

$$\text{div}\,\mathbf{V} = \sum_i \left(\frac{\partial}{\partial q_i}\dot{q}_i + \frac{\partial}{\partial p_i}\dot{p}_i\right) = 0 \tag{5}$$

which is an equation that, we know, expresses the fact that the fluid is non-compressible, that is, that the same number of representative points always occupies the same volume of extension-in-phase. The continuity equation then gives

$$\frac{\partial \sigma}{\partial t} + \mathbf{V} \cdot \text{grad}\,\sigma = 0 \qquad \text{or} \quad \frac{\mathrm{D}\sigma}{\mathrm{D}t} = 0 \tag{6}$$

$\mathrm{D}/\mathrm{D}t$ being the derivative taken along the motion of the representative points. Equation (6) signifies that by following the motion of the representative points, the density σ remain constant—which is another way of symbolizing the property of non-compressibility. These results make up Liouville's theorem, which may be expressed by saying that if, at an initial instant a collection of representative points occupies a volume $d\tau_0$ of extension-in-phase, at every later instant it will occupy a volume $d\tau = d\tau_0$.

The equation

$$\frac{\mathrm{D}\sigma}{\mathrm{D}t} = \frac{\partial \sigma}{\partial t} + \sum_i \left(\frac{\partial \mathrm{H}}{\partial p_i}\frac{\partial \sigma}{\partial q_i} - \frac{\partial \mathrm{H}}{\partial q_i}\frac{\partial \sigma}{\partial p_i}\right) = 0 \tag{7}$$

being of the first order with respect to time, its solution $\sigma(q_i, p_i, t)$ is determined if we know the initial form $\sigma(q_i, p_i, t_0)$; moreover, this equation obviously allows the solution $\sigma = \text{const.}$

In Classical Statistical Mechanics it is customary, since the time of Boltzmann and Gibbs, to consider ensembles of systems represented by clouds of points in extension-in-phase and to assume as a statistical postulate that the solution σ = const. is the one that must be chosen. In other words, one assumes—as the Liouville theorem makes it possible to do—that the probability of finding the representative point of a system in an element of volume $d\tau$ of extension-in-phase is proportional to the magnitude of $d\tau$. In the frequently occurring case in which the systems are conservative, that is, have a constant energy E (known within dE), one considers in the extension-in-phase a layer having a thickness corresponding to dE and covering the multiplicity of $6N-1$ dimensions defined by E = const. The statistical postulate we have referred to must be applied inside this layer. The elements $d\tau$ of the layer have an expression of the form πdE, and π can then serve as a measure of the probability of presence of a representative point in $d\tau$. Due to the Boltzmann relation $S = k \ln P$ between the entropy S and the probability $P = \pi$, one can define the entropy of the ensemble of the systems.[1]

One can try to justify the arbitrary choice of the solution σ = const. of (6) by introducing the "ergodic hypothesis" according to which every trajectory of the representative point of the system, ultimately, would end up by covering, without any duplication, all of that portion of extension-in-phase accessible to it (for example, the entire layer between E and E + dE in the case of conservative systems). But that hypothesis is obviously over-rigorous and, more especially, is completely inadequate in the case of periodic motions. So a considerably modified version has been substituted for it, usually called "the quasi-ergodic hypothesis", according to which every representative point of the system, save for the exceptional cases of disappearing probability, will finally pass infinitely close to every point of the extension-in-phase region accessible to that trajectory.

An all-important consequence of either of these hypotheses is this: The average in time $\lim_{T\to\infty} 1/T \int_t^{t+T} dt\, A$, where A is a magnitude associated with the system, is independent of t and is equal to the average

$$\overline{A} = \frac{1}{\Omega}\int_{\Omega} d\tau\, A \tag{8}$$

of the quantity A taken over the volume Ω of extension-in-phase when

[1] See, for example, H. A. Lorentz, *Les Théories statistiques en thermodynamique*, B. G. Teubner, Leipzig, 1916, and Francis Perrin, *Mécanique statistique quantique*, Gauthier-Villars, Paris, 1939.

we assume the probabilities proportional to $d\tau$. This equality of the average in time and average (8) is very important to the coherence of the theory. Unfortunately, the ergodic hypothesis, even in its modified quasi-ergodic form, is very hard to justify. In order to do it, it seems necessary to introduce in one form or another a postulate of a statistical nature (for example, Boltzmann's hypothesis of molecular chaos)—a kind of hypothesis that is, by its very nature, foreign to the laws of Classical Mechanics and not justifiable by it.

To sum up: Among all the possible solutions of equation (6) resulting from Liouville's theorem, Classical Statistical Mechanics arbitrarily chooses the simplest $\sigma = \text{const.}$ and tries to justify this choice by the demonstration of an ergodicity theorem—a demonstration that requires the introduction of a statistical postulate foreign to the laws of Classical Dynamics.

But let us now go back to the interpretation of Wave Mechanics by the causal theory and to the statistical interpretation of $|\Psi|^2$. Here, in the case of single particles as well as of systems, we consider a space formed by means of coördinates of type q_i without introduction of the conjugate momenta p_i as required by extension-in-phase, and for a group of systems a density $\rho(q_i, t)$ is defined obeying the continuity equation (1) where $\mathbf{v}$ is the velocity in physical space or in configuration space given by the guidance formula $\mathbf{v} = -1/m \text{ grad } \varphi$. Except for the special cases where $\text{div } \mathbf{v} \sim \Delta\varphi$ is zero, this equation does not reduce to $D\rho/Dt = 0$. The problem is thus decidedly different from the one arising in Classical Statistical Mechanics. However, here again we have arbitrarily adopted a particular solution of the equation for ρ, namely $\rho = a^2(q_i, t) = |\Psi|^2$ which, in the case of a plane monochromatic wave, reduces to $\rho = \text{const.}$ This hypothesis was suggested to us by the obvious *a priori* equivalence, in the case of a uniform rectilinear motion in a given direction, of all parallel trajectories and of all positions on these trajectories. One may think that, as in Classical Statistical Mechanics, it is necessary to try and justify this natural but arbitrary hypothesis by the demonstration of a sort of ergodicity theorem.

In Classical Statistical Mechanics we try to justify the hypothesis $\sigma = \text{const.}$ in the following manner: We consider an aggregate of systems for which the density σ in extension-in-phase has an initial form $\sigma(q_i, p_i, t_0)$, and we assume—and this is the purely statistical element that is added to the laws of Dynamics—that this aggregate of systems is subjected to entirely uncoördinated perturbations. We seek to deduce that σ tends towards the form $\sigma = \text{const.}$, whatever its initial form may be, so that everything takes place as if the ergodic hypothesis were correct. A similar path was followed by David Bohm in a recent paper [8]. Within the framework of the causal theory, he considers an ensemble of systems whose representative points in configuration space have initially any density $\rho(q_i, t_0)$, and he introduces the statistical hypothesis that these systems are subjected to successive collisions, with the parameters defining each collision having values distributed at random. From this he deduces that, whatever its initial form may be, $\rho(q_i, t)$ tends toward $a^2(q_i, t) = |\Psi|^2$.

3. A brief summary of Bohm's paper of January 1953

Bohm began by explaining the similarity of the question to the theorem of ergodicity in Classical Statistical Mechanics, putting

$$\rho(q_i, t) = f(q_i, t)|\Psi(q_i, t)|^2 = f(q_i, t)a^2(q_i, t) \tag{9}$$

which is always permissible, and by pointing out that since the equations of the causal theory furnish the relation $\partial a^2/\partial t + \operatorname{div}(a^2 \mathbf{v}) = 0$ and since the conservation of the particles requires the continuity relation (1), we conclude

$$|\Psi|^2 \frac{\partial f}{\partial t} + f \frac{\partial |\Psi|^2}{\partial t} + f \operatorname{div}(|\Psi|^2 \mathbf{v}) + |\Psi|^2 \mathbf{v} \cdot \operatorname{grad} f = 0 \tag{10}$$

and subsequently

$$\frac{\mathrm{D}f}{\mathrm{D}t} = \frac{\partial f}{\partial t} + \mathbf{v} \cdot \operatorname{grad} f = 0. \tag{11}$$

So it is the function f that here obeys the same relation as the density σ does in extension-in-phase, that is, remains constant when we follow the motion of the particles. The theorem analogous to the ergodicity theorem will thus consist of demonstrating that the f function tends to take on a constant value if a statistical hypothesis of randomness is assumed.

I will content myself with a rapid summary of the trend of Bohm's reasoning, referring the reader to the text of his article for the details of calculation. Bohm considers a hydrogen atom that is in a doubly degenerate state of excitation. The wave function in cylindrical coördinates then has the form

$$\Psi = \sqrt{2}\, g(\rho, z)(c_1 \cos\varphi + c_2 \sin\varphi) e^{\frac{i}{\hbar} E_0 t} \tag{12}$$

with

$$|c_1|^2 + |c_2|^2 = \frac{1}{2\pi} \qquad \text{and} \int_{-\infty}^{\infty} dz \int_0^{\infty} d\rho\, \rho[g(\rho, z)]^2 = 1. \qquad (13)$$

The particle's motion is easily deduced from this by means of the guidance formula. Bohm then supposes that the atom is subjected to a collision with a very heavy particle whose motion may be classically described, and he applies the perturbation method in calculating the variations of c_1 and c_2. The values of these quantities before the collision are of the form $c'_j(c_1, c_2, \mathbf{h}, \mathbf{u})$, where $\mathbf{u}$ is the initial velocity of approach of the incident particle, $\mathbf{h}$ the vector defined by dropping a perpendicular from the center of the atom onto the initial direction of the particle's motion ($\mathbf{u}$), c_1 and c_2 being the values after the collision. $\mathbf{h}$ and $\mathbf{u}$ are thus the "impact parameters" that define the collision.

It is at this point that Bohm introduces a statistical postulate by assuming that the various atoms in an aggregate of hydrogen atoms undergo collisions corresponding to values distributed at random of $\mathbf{h}$ and $\mathbf{u}$ so that, if the c_j are the same for all the atoms at the outset, they are found to be statistically distributed after the collisions.

Proceeding from this statistical postulate, Bohm performs calculations that I will not go into here and that undoubtedly would require revision at certain points. But he manages in this way to show that the function $f(q_i, t)$ must tend toward a constant value when t tends toward infinity. Now, if the probability of presence $\rho(q_i, t)$ is normalized to one by $\int d\tau\, \rho = 1$, and if the Ψ wave function is also normalized to one by $\int d\tau\, |\Psi|^2 = 1$, the definition of f by (9) has as a consequence, if f is a constant, that this constant is equal to 1. So f must tend toward 1 and, consequently ρ must tend toward $|\Psi|^2$. The sought-for demonstration seems thus to be achieved, but in a rather special case, since it was limited to a hydrogen atom in a doubly degenerate state of excitation.

Whatever criticism may be levelled at the rigor of Bohm's reasoning and at its lack of generality, it does appear that we are here confronted with a problem very similar to that which the ergodicity demonstation seeks to solve in Classical Statistical Mechanics. Just as we sought to show how the introduction of perfectly random perturbations tends to produce the density $\sigma =$ const. of the representative points in extension-in-phase, we must here seek to show that the action of perfectly random perturbation must make the function $f = \rho/|\Psi|^2$ tend toward a constant which, by reason of the normalization of ρ and of $|\Psi|^2$, will have to be equal to 1.

To sum up: In regard to the justification of the statistical significance of $|\Psi|^2$ in the causal theory, the situation does not seem to me any worse than it is in regard to the rigorous justification of Classical Statistical Mechanics, and that conclusion strikes me as rather encouraging.

4. Supplementary observations

The principle of Bohm's demonstration consists of pointing out that the Ψ wave of a particle or of a system is always slightly perturbed by the existence of very small external actions (weak collisions, for example) and of assuming that these small perturbing potentials represent these actions with their entirely random fluctuations. It is here a question of potentials of the classical type, but Vigier has quite correctly pointed out that one could consider small perturbing quantum potentials resulting from small random fluctuations in boundary conditions (for example, the thermal motion of the walls of a receptacle). In seeking to answer one of Einstein's objections to the guidance formula, we have already seen (Chapter XI, section 3) that there may be a justification for introducing such fluctuations in boundary conditions.

It seems to me that Vigier's idea may be summarized as follows: To a given form of the Ψ wave there corresponds a congruence L formed by an infinite number of possible trajectories defined from the phase of the Ψ wave by means of the guidance formula. Small external actions of the classical type or small fluctuations of boundary conditions manifested as small perturbations of the quantum potential cause the particle to jump constantly from one trajectory to another of the L congruence. If these jumps are completely random, it is clear that the particles (or the points representative of systems) are going to be agitated by a sort of Brownian movement, with the curves of the L congruence continuing to define the average trajectories. One would then have to demonstrate that this jumping about from one trajectory to another achieves on the average the probability of presence $|\Psi|^2$. Such a demonstration would probably require use of the properties of "Markov chains". [2]

When we studied certain consequences of the guidance formula, we

[2] Bohm and Vigier have recently published a justification of the statistical significance of $|\Psi|^2$ based on the sort of conception put forward here. [16].

saw that in certain quantized states the motion of the particle predicted by the guidance formula may be so simple (for example, immobility or uniform circular motion) that one cannot at all see how the distribution of the density of the probability of presence in $|\Psi|^2$ may be realized. But small random perturbations will transform Ψ into $\Psi + \delta\Psi$, where $\delta\Psi$ is a very small modification of Ψ. This very small modification, by modifying the wave's phase φ at random, is sufficient to transform the unperturbed simple motion into a very complicated one of a quasi-Brownian type. If one *assumes* that this quasi-Brownian movement fulfills the distribution of probability of presence $|\Psi + \delta\Psi|^2 \simeq |\Psi|^2$, one understands why—if we leave aside the inevitable random perturbations—one is led to consider the density of the probability of presence in the unperturbed Ψ state as equal to $|\Psi|^2$. The connection between this previously established result with the ideas that we have just developed here is obvious.

I think the foregoing may be summarized in the following manner: If we except random perturbations due to external actions or to random fluctuations of boundary conditions, the motion of the particle (or of the representative point) is given directly from the phase of the unperturbed Ψ wave by the guidance formula. But, in reality, small random external actions and small random fluctuations in boundary conditions always do enter into the picture, and these actions which, by producing a sort of Brownian movement that causes the particle (or representative point) to jump constantly from one unperturbed motion to another, assures realization of the density of presence as $|\Psi|^2$. In this way, while still retaining the average physical significance of the trajectories predicted by the causal theory, one succeeds in superimposing upon them a kind of Brownian movement. It is curious to note that in this way there would be achieved a synthesis of the conceptions of the causal theory and of Einstein's frequently reiterated affirmation that the successes of the statistical interpretation of Wave Mechanics imply underlying particle-movements of a Brownian character.[3]

[3] Since 1954, when this passage was written, I have come to support wholeheartedly an hypothesis proposed by Bohm and Vigier. According to this hypothesis, the random perturbations to which the particle would be constantly subjected, and which would have the probability of presence in terms of $|\Psi|^2$, arise from the interaction of the particle with a "subquantic medium" which escapes our observations and is entirely chaotic, and which is everywhere present in what we call "empty space".

Chapter XIV

PAULI'S OBJECTIONS TO THE THEORY OF THE PILOT-WAVE

1. The discussion of the pilot-wave theory at the Solvay Congress of October 1927

In the conclusion of my paper in the *Journal de Physique* of May 1927, after I had set forth the main points of the theory of the Double Solution, I made the following observation: The theory of the Double Solution had led me to the guidance formula

$$\mathbf{v} = -c^2 \frac{\operatorname{grad} \varphi + \frac{e}{c}\mathbf{A}}{\frac{\partial \varphi}{\partial t} - eV} \tag{1}$$

where the point of departure is Relativistic Mechanics utilizing one Ψ function—the only form of Relativistic Mechanics then known; and then I was led to the formula for the corresponding probability of presence

$$\rho = \text{const. } a^2 \left(\frac{\partial \varphi}{\partial t} - eV\right). \tag{2}$$

The theory of the Double Solution had also led me to introduce the quantum potential defined by

$$Q = \frac{\hbar^2}{2m}\frac{\Box f}{f} = \frac{\hbar^2}{2m}\frac{\Box a}{a} \tag{3}$$

for the Newtonian approximation.

It then struck me that one could express all these quantities by the sole use of the amplitude a and the phase φ of the wave written in the form $ae^{\frac{i}{\hbar}\varphi}$, and I wrote at the end of my paper, "If one does not care to have recourse to the principle of the Double Solution, it might be possible to adopt the following point of view: The existence of both the particle and the continuous Ψ wave will be assumed as two distinct realities, and a postulate will be adopted according to which the motion of the particle is determined as a function of the phase (of the Ψ wave) by means of the guidance formula. In that way we imagine the Ψ wave guiding the particle so that it becomes a pilot-wave."

But I hastened to add, "by thus taking the guidance equation as a postulate, one avoids having to justify it by the principle of the Double Solution—but that can only be a provisional attitude. It will very likely be necessary to *reincorporate* the particle into the wave phenomenon, and one will thus probably be led back to ideas similar to those developed in the foregoing considerations."

These quotations indicate very clearly what my feeling was at the end of that work. I considered the pilot-wave point of view as being of practical use but also as really justifiable only within the framework of a more comprehensive theory of the "double-solution" type.

However, when I was asked to present a paper on Wave Mechanics at the Solvay Congress that was to be held in Brussels during October 1927, I balked at the difficulties of mathematically justifying the double-solution point of view, and I contented myself with a presentation of the pilot-wave point of view. At the Solvay Congress, while a few of the "old guard" (Lorentz, Einstein, Langevin, Schrödinger) insisted on the necessity of finding a causal interpretation of Wave Mechanics—without, however, coming out in favor of my efforts—Bohr and Born, along with their young disciples (Heisenberg, Dirac, etc.), came out categorically in favor of the new purely probabilistic interpretation that they had developed, and they did not even discuss my point of view. Pauli was the only one to present a definite objection to my theory, and he did so by examining the case of a collision between a particle and a rotator, which Fermi had studied a short while before.

I will explain Pauli's objection, but in order to do that, I must first review the main lines of argument presented by Fermi.

2. The collision of a particle and a plane rotator according to Fermi

Let us first recall that what we call a "plane rotator" is a point-mass (mass M) subject to displacement in a single plane and remaining at a fixed distance R from the origin of its coordinates. From the mechanical point of view, this rotator is characterized by its moment of inertia $I = MR^2$.

The rotator, whose dimensions we neglect, is assumed to be situated at a point 0 chosen as the origin of the coördinates; and a particle moving in a straight line along the z-axis collides with the rotator.[1] An interaction, a collision in the broad sense of the word, takes place between the particle and the rotator. If z is the abscissa of the particle and ϑ the polar angle that fixes the configuration of the rotator, the potential function corresponding to the interaction will be of the form $U(z, \vartheta)$; it will not differ from zero except for very small values of z (positive or negative) and will be periodic, with the period 2π in ϑ.

Granting all this, the problem to be solved is: If we know that at the beginning of the motion the particle, whose mass is m, has a certain velocity v_0, determine the various possible results of the collision between the particle and the rotator.

The system will be represented by considering the configuration space formed by the two variables z and ϑ. The equation of propagation in this configuration space will be

$$\frac{1}{m}\frac{\partial^2 \Psi}{\partial z^2} + \frac{1}{MR^2}\frac{\partial^2 \Psi}{\partial \vartheta^2} + \frac{2}{\hbar^2}[E - U(z, \vartheta)]\Psi = 0 \tag{4}$$

E being the total energy of the system. Let us make the substitution of variables

$$\xi = \sqrt{I}\,\vartheta \qquad \zeta = \sqrt{m}\,z. \tag{5}$$

The equation of propagation becomes

$$\frac{\partial^2 \Psi}{\partial \xi^2} + \frac{\partial^2 \Psi}{\partial \zeta^2} + \frac{2}{\hbar^2}[E - U(\xi, \zeta)]\Psi = 0 \tag{6}$$

$U(\xi, \zeta)$ being zero whenever ζ is not very small and being periodic with the period $2\pi\sqrt{I}$ in ξ.

The frequency of the Ψ wave being $\nu = E/h$, we can write

$$\frac{2}{\hbar^2}E\Psi = -\frac{1}{V^2}\frac{\partial^2 \Psi}{\partial t^2} \qquad \text{with} \qquad V = \sqrt{\frac{E}{2}} \tag{7}$$

and the equation of the waves becomes

$$\frac{\partial^2 \Psi}{\partial \xi^2} + \frac{\partial^2 \Psi}{\partial \zeta^2} - \frac{1}{V^2}\frac{\partial^2 \Psi}{\partial t^2} = \frac{2}{\hbar^2}U(\xi, \zeta)\Psi. \tag{8}$$

Now here is the very ingenious observation that made it possible for Fermi to solve the problem very simply: Since the function U differs from zero only in the immediate vicinity of the ξ axis, where ζ is very

[1] In his explanation, Fermi had the particle remain only in a plane $x\,O\,y$.

small and U is periodic with period $2\pi\sqrt{I}$ in ξ, everything takes place as if the ξ axis played in configuration space, of two dimensions ξ, ζ, the role of a grating with equidistant intervals $2\pi\sqrt{I}$, which diffracts the incident Ψ wave.

Let us first construct this incident Ψ wave. The Ψ_1 wave that represents the initial state of the incident particle before interaction is simply

$$\Psi_1 = a_1 e^{\frac{i}{\hbar}[E_1 t - mv_0 x]} = a_1 e^{\frac{i}{\hbar}[E_1 t - \sqrt{2E_1}\,\zeta]} \tag{9}$$

with

$$E_1 = \tfrac{1}{2}mv_0^2.$$

Then let ω_0 be the initial angular velocity of the rotator. The initial wave Ψ_2 of the rotator is

$$\Psi_2 = a_2 e^{\frac{i}{\hbar}[E_2 t - I\omega_0\theta]} = a_2 e^{\frac{i}{\hbar}[E_2 t - \sqrt{2E_2}\,\xi]} \tag{10}$$

with

$$E_2 = \tfrac{1}{2}I\omega_0^2.$$

In fact, Ψ_2 must be a solution of the equation of the isolated rotator

$$\frac{\partial^2 \Psi_2}{\partial \zeta^2} + \frac{2E}{\hbar^2}\Psi_2 = 0 \qquad \text{for } E = E_2. \tag{11}$$

Since Ψ_2 must take on the same value when ξ increases by $2\pi\sqrt{I}$, one must have

$$\frac{1}{\hbar}\sqrt{2E_2}\,2\pi\sqrt{I} = 2\pi n \qquad (n \text{ integral}) \tag{12}$$

that is

$$E_2 = \tfrac{1}{2}I\omega_0^2 = \frac{n^2\hbar^2}{2I}. \tag{13}$$

This is the well known formula that gives the quantized values of the energy of the plane rotator. The initial state of the rotator being of necessity quantized, the initial angular velocity ω_0, of the rotator obeys relation (13).

The Ψ wave of the particle-plus-rotator system in the configuration space has as its initial form the product $\Psi_1\Psi_2$, namely

$$\Psi = a_1 a_2 e^{\frac{i}{\hbar}[(E_1+E_2)t - \sqrt{2E_1}\,\zeta - \sqrt{2E_2}\,\xi]}. \tag{14}$$

Let us put

$$\cos\alpha = \sqrt{\frac{E_2}{E}}, \quad \sin\alpha = \sqrt{\frac{E_1}{E}}, \quad \lambda = \frac{V}{\nu} = \frac{h}{\sqrt{2E}} \tag{15}$$

where $E = E_1 + E_2$. This gives us for the initial form of Ψ

$$\Psi = ae^{2\pi i\left[\frac{E}{h}t - \frac{\zeta\sin\alpha + \xi\cos\alpha}{\lambda}\right]}. \tag{16}$$

In configuration space the incident Ψ wave is then a plane monochromatic wave of frequency $\nu = E/h$ and wave-length λ whose normal makes an angle α with the ξ axis.

One can then, by making use of Fermi's observation, consider the interaction of the rotator and the particle as producing diffraction of the Ψ wave in configuration space. The directions of maximum intensity of the diffracted wave are given by the relation expressing the coincidence of phases for a linear grating of period $2\pi\sqrt{I}$ along the ξ axis, namely

$$2\pi\sqrt{I}(\cos\alpha' - \cos\alpha) = k\lambda \qquad (k \text{ integral}) \tag{17}$$

as is easily seen from Fig. 8.

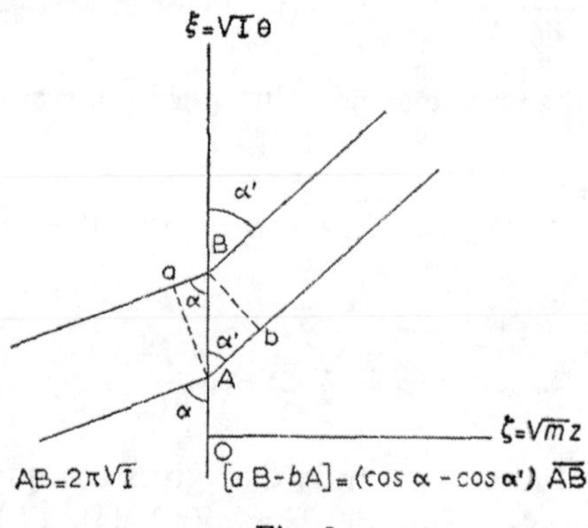

Fig. 8

After the interaction, the Ψ wave will then have the form

$$\Psi = \sum_{\alpha'} c_{\alpha'} e^{2\pi i\left[\frac{E}{h}t - \frac{\zeta\sin\alpha' + \xi\cos\alpha'}{\lambda}\right]}, \tag{18}$$

α' being able to take on all the discrete values given by (17), means that the incident plane wave is thus transformed by the interaction into a superposition of diffracted plane waves.

Let us now return to the variables z and ϑ and let us replace λ by its value; we find the expression of the final Ψ to be

$$\Psi = \sum_{\alpha'} c_{\alpha'} e^{\frac{i}{\hbar}[Et - \sqrt{2mE}\, x \sin \alpha' - \sqrt{2IE}\, \vartheta \cos \alpha']}. \tag{19}$$

According to the principles of Wave Mechanics, if the final state of motion of the particle and the rotator is determined by observation, one will find a state of motion represented by one of the plane waves of development (19). Now one of these plane waves represents a motion with energy E_1' and momentum $\sqrt{2mE_1'} = \sin \alpha' \sqrt{2mE}$, to which the particle velocity $\sqrt{2E/m} \sin \alpha'$ corresponds. This motion has an energy $E_1' = E \sin^2 \alpha'$. To this same plane wave in configuration space there corresponds a correlated motion of the rotator with energy E_2' and momentum $\sqrt{2IE_2'} = \cos \alpha' \sqrt{2IE}$, which gives $E_2' = E \cos^2 \alpha'$. One has thus: $E = E_2' + E_1'$, which expresses the conservation of energy. There is no conservation of momentum because we have implicitly assumed that the rotator is fixed at the time of the collision, and because the particle is forced to remain on the axis.

Taking definition (15) of λ into account, equation (17) gives us

$$\cos \alpha' = \cos \alpha + \frac{1}{\sqrt{2IE}} k\hbar \tag{20}$$

and since

$$\cos \alpha = \sqrt{\frac{E_2}{E}} = \frac{1}{\sqrt{2IE}} n\hbar \tag{21}$$

it follows that

$$\sqrt{E} \cos \alpha' = \frac{1}{\sqrt{2I}} (n + k)\hbar \tag{22}$$

whence finally

$$E_2' = E \cos^2 \alpha' = \frac{1}{2I} (n + k)^2 \hbar^2. \tag{23}$$

n and $n + k$ being positive or negative integers or zero.

If then, by observation, we determine the states of motion after the collision, we will always find that the effect of the collision has been to make the rotator pass from the initial quantized state characterized by the quantum number n to a final quantized state characterized by the quantum number $n + k$, so that the rotator will always be, finally as well as initially, in a quantized state.

Three cases may then arise:

1. $k = 0$. — The particle and the rotator retain their initial states. The collision takes place without any exchange of energy; it is "elastic".

2. $k > 0$. — The particle gives up energy to the rotator, which make a transition into a quantized state of energy higher than the energy of the initial quantized state. This is an "inelastic collision of the first kind".

3. $k < 0$. — The rotator makes a transition from its initial quantized state to a quantized state of lesser energy, transferring energy to the incident particle. Using an expression introduced recently by Klein and Rosseland, we may say that there is then an "inelastic collision of the second kind".

This is the beautiful calculation made in 1927 by Fermi describing the collision of a particle and a plane rotator.

3. Pauli's objection to the guidance formula

As I recalled previously, at the Solvay Physical Congress of October 1927, Pauli utilized Fermi's calculation in criticizing the guidance formula, which is basic in the theory of the Double Solution and also in the pilot-wave theory.[2]

Formula (19) gives us the final form of the Ψ of the particle-rotator system in configuration space after the collision, the form being that of a superposition of plane monochromatic waves; and the probabilistic interpretation of the Ψ tells us that each of these plane waves corresponds to a state of motion of the particle and to a state of motion of the rotator, the two states being correlated with each other. There is no difficulty in that, Pauli said; but a very serious difficulty arises if one wishes to attribute to the representative point of the system in configuration space z, ϑ a motion defined by the guidance formula. Indeed, we would then have to write the final Ψ (19) in the form $\Psi = ae^{\frac{i}{\hbar}\varphi}$ with a and φ real and, by reason of the superposition of the plane monochromatic waves, the phase φ would then have a very complicated form; the final motion of the representative point in configuration space would then be a very complex motion without any relation to the final quantized states of the rotator which the orthodox inter-

[2] See *Électrons et photons, Comptes rendus du Ve Conseil de Physique Solvay*, Gauthier-Villars, Paris, 1928, pp. 282—284.

pretation causes immediately to appear and the existence of which is confirmed by experiment. Pauli added that it was not possible to get rid of the difficulty by introducing a limited Ψ wave since, by reason of the period 2π of the coördinate ϑ, it is impossible to assume that the Ψ wave is limited in the direction $O\vartheta$ (ξ axis).

The latter argument seems to me to prove, above all else, the wholly fictitious character of configuration space and of wave-propagations in that space. (How could one, in fact, attribute a physical meaning to the $O\vartheta$ axis along which we indefinitely assign the values of an angle that, in reality, varies only from O to 2π?) The argument dit not convince me, and I saw in the limiting of the wave-trains a way to avoid Pauli's objection. At the time I answered him thus: "The difficulty pointed out by Pauli has its analogue in Classical Optics. One cannot speak of a beam diffracted by a grating in a given direction except when the grating and the incident wave are laterally limited, for otherwise the diffracted beams would overlap and be obliterated in the incident wave. In Fermi's problem there must also be a Ψ wave limited laterally in configuration space."

Let us examine the question more closely. One may grant Pauli that the incident Ψ wave is not limited in the direction $O\vartheta$, but it is necessarily limited in the direction Oz, since the Ψ_1 wave of the incident particle is necessarily a wave-train that we assume to be almost monochromatic, but which is nevertheless of finite length. The result is that the final state *is* represented by formula (19), but with the important reservation that each of the plane monochromatic waves occurring in this development is in reality an almost monochromatic wave-train whose dimensions are limited in the direction Oz. Each of these wave-trains corresponds to a different value of the quantized angle α' and, consequently, moves with an overall velocity along Oz, as we have seen, equal to $\sqrt{2E/m} \sin \alpha'$. Since the component velocities of the various diffracted wave-trains are not the same along Oz, on leaving the grating constituted by the $O\vartheta$ axis the components will finally separate, and though remaining indefinite in the direction $O\vartheta$, will correspond to the separate regions of the Oz axis.

At the end of the collision the motion of the representative point, which we assume to be given by the guidance formula, will shift the point *into one or the other* of the separated emergent wave-trains, and which one will depend on the initial position of the point in the incident wave-train. The guidance formula applied to the final position of the representative point will then show us that the rotator is in a final quantized state and that the particle possesses the motion correlated to the quantized state. In this way we will find precisely Fermi's conclusions, in agreement with the experimental existence of only the quantized states of the rotators.

If, instead of forcing the particle to remain along a straight line Oz we merely made it—as Fermi did in his paper—remain in a plane xOy, there would be occasion to limit the incident wave-train in both the x and y directions, and we would still have a final separation of wave-trains after the interaction.

So I had perceived, as the quotation given above shows, that the answer to Pauli's objection would have to rely on the fact that the wave-trains are always limited. And this idea has been taken up again by Bohm in his recent Papers.

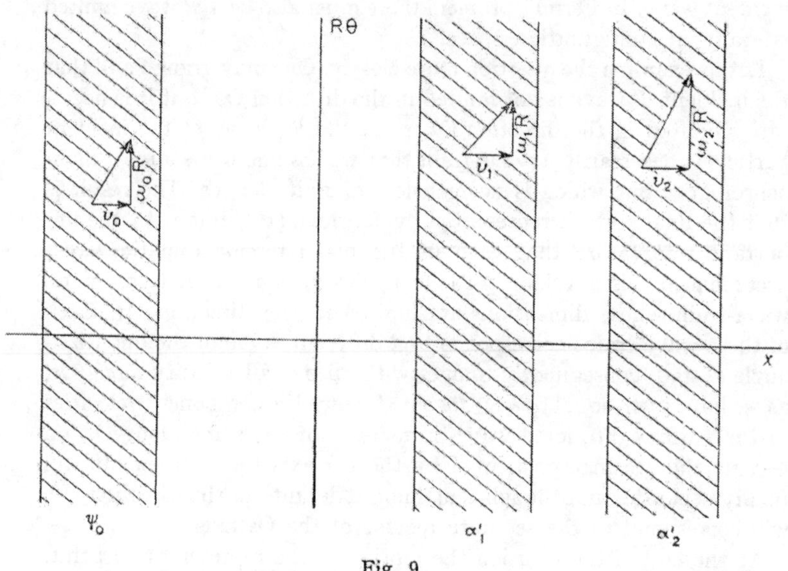

Fig. 9

So we have here for the first time run into the idea, which must never be forgotten—as Schrödinger quite recently again emphasized—that wave-trains are always limited. It is a point that is most often simply passed over in silence, both in Classical Optics and in Wave Mechanics. We will see, however, that for certain problems, the limitation of wave-trains is of capital importance.

4. The abandonment of attempts at a causal interpretation of Wave Mechanics after 1927

In the months that followed the Solvay Congress of October 1927, I abandoned the pilot-wave approach that I had maintained. But not because of Pauli's objection. For, as I say, I thought I had found the way in which to overcome it. Rather, I abandoned the pilot-wave approach for other, more general, reasons, which I put forward notably in my first course at the Institut Henri Poincaré, subsequently published under the title of *Introduction à l'étude de la Mécanique ondulatoire.*[3] A summary of those reasons follows.

The particle, conceived as a physical reality, cannot, I said to myself, be guided by the Ψ wave, whose probability-representational character (at once subjective and conditioned by the knowledge of its user) had been plainly revealed by the development of Wave Mechanics.

This fictitious character of the Ψ wave was already forced upon us for the Ψ wave associated with only a single particle in ordinary space. It thus became all the more inescapable in the case of the Ψ wave of a system, which is propagated in the system's configuration space, which is purely abstract. Besides, even in the case of a single particle—a case in which the wave $\Psi(x, y, z, t)$ seems to be expressed in terms of the three ordinary coördinates of physical space—the variables x, y, z in the accepted interpretation do not really represent the coördinates of any point in physical space, but rather the possible positions of the particle in that space. This distinction, subtle in appearance, is none the less necessary for the clear understanding of how one passes from the general case of a system of particles to the case of a single particle. Thus, even in this latter case where three coordinates x, y, z suffice, the wave function $\Psi(x, y, z, t)$ cannot represent a *field* having physical reality in three-dimensional physical space.

Heisenberg, moreover, had insisted on the fact that every experimental observation of a particle or system instantly modifies the form of the Ψ wave by obliging us to effect what he called "the reduction of the probability packet". By the effect of experimental observation alone, a whole section of the Ψ wave disappears "as does the expectation of a possible occurrence when that occurrence fails to materialize". These properties of the Ψ wave, which are the ones involved in Schrödinger's and in Einstein's objections, are comprehensible if this wave

[3] Hermann, Paris, 1930.

is merely a representation of probabilities. But they do not permit us to attribute to it the role of a physical reality. So it cannot be a physical agent guiding the motion of a particle.

On close examination one also sees that, if the particle's motion were guided by the Ψ wave, it would depend not only on the position that the particle had originally, but on all the positions it might have occupied, *but actually did not occupy.* This paradoxical set of circumstances excludes all possible hope of any real return to a classical causal conception of the particle's motion solely by means of the pilot-wave concept.

All these basic objections are set forth in the work that I cited above. And still other objections may be found. Let us mention, for example, the following one due to Francis Perrin—an objection we will have occasion to consider later at much greater length. If we consider the divergent spherical wave that represents the isotropic emission of particles by a source, this wave is expressed by $(1/r)e^{-ikr}$, and its amplitude decreases as the inverse of the distance from the source. The $|\Psi|^2$ then decreases in inverse proportion to the square of the distance from the source—which corresponds to the interpretation of this quantity as a density of probability of presence. At a very great distance from the source, the Ψ wave thus has only a very small amplitude, and yet, if it enters an interference apparatus, the theory of the pilot-wave tells us that the particle's motion in that apparatus is guided by the propagation of a Ψ wave whose amplitude is infinitesimal—and that is rather disturbing from the physical point of view. Of course, from the purely mathematical viewpoint one could reply that the quantum potential acting on the particle is proportional to $\Delta a/a$ and consequently does not depend on the absolute value of a; but physically, if the Ψ wave were really a physical agent acting on the particle, it is difficult to see how this action could be the same, no matter how small the amplitude of the Ψ wave is. Perrin's example shows very clearly that the Ψ wave, precisely because it represents a probability of presence that necessarily diminishes as the spherical wave spreads out in space, cannot be considered a physical agent guiding the particle.

Such were the considerations that led me, in 1928, to abandon the pilot-wave theory as untenable. The original form of my ideas, *i.e.* the theory of the Double Solution, did not seem to me to run into the same difficulties, but I had become convinced that its mathematical justification, if it were possible, was beyond my capacities. And when I was appointed to a teaching post on the Faculté des Sciences de Paris as of November 1928, I did not feel I had the right to teach a viewpoint that I was not in a position to justify. So I joined the ranks of the purely probabilistic interpretation put forward by Born, Bohr and Heisenberg. In spite of the criticism of a few isolated scientists such as Einstein and Schrödinger, the purely probabilistic interpretation was subsequently adopted by almost all theoretical physicists. And it likewise became the "orthodox" interpretation of Wave Mechanics. To my knowledge, from 1928 to 1951 no serious attempt was made to effectively construct any other interpretation more closely approaching classical conceptions.

And then in January 1952 appeared the two Papers by David Bohm: We will now analyze the main arguments of these two Papers.

Chapter XV

BOHM'S THEORY OF MEASUREMENT AND THE STATISTICAL SCHEMA OF THE CAUSAL THEORY

1. Bohm's papers of January 1952

The two articles published simultaneously by David Bohm in January 1952 in *The Physical Review* [3] once more focussed attention on the question of the interpretation of Wave Mechanics. In these papers Bohm reverts to the pilot-wave theory in the form I had given to it at the 1927 Solvay Congress. He assumes that the Ψ wave is a physical reality (even the Ψ wave in configuration space!). I have already stated why such an hypothesis appeared absolutely untenable to me.

Bohm writes (p. 177, column 2) that, after reduction of the packet of probability by an observation, "the wave function can . . . be normalized because the multiplication of Ψ by a constant changes no physically significant quantity, such as the particle velocity or the 'quantum mechanical' potential". It is perfectly true that, if the Ψ function is multiplied by a constant, neither the gradient of its phase φ nor the relation $\Box a/a$ is modified; but it is obvious that, if Ψ represented a physical reality, one would have no right at all to multiply its amplitude by a constant. What permits us to do that is the very fact that Ψ is merely a representation of probability and that the only object in normalizing it is to be able to express in absolute value the probability of presence in terms of the expression $|\Psi|^2$. Moreover, the reduction of the packet does not involve merely a renormalization of the Ψ function, but in general a sudden and complete modification of its form also, and this would be incomprehensible if Ψ were a physical reality.

Bohm's papers contain still other statements that strike us as dubious. For example, he is undoubtedly right in saying that on a very reduced scale (10^{-13} cm or less) the guidance formula, and consequently the statistical significance of the Ψ, could very well no longer hold, but the modification of the equation of propagation which he proposes as a remedy seems to me artificial.

Nevertheless, if Bohm's work calls for certain reservations, it also has merits that seem to me unquestionable. In particular, he has once more focussed attention on the possibility of an interpretation of Wave Mechanics different from the one that is now prevalent, and he has shown that it is not pointless to submit the whole question to a painstaking re-examination. The problem posed is clearly defined in the first three sections of his first paper, and several pertinent and interesting remarks appearing in those sections deserve mention.

Bohm has come up with the same results that I formerly obtained relative to the particle's motion in a quantized atom, especially as they concern the particle's immobility in the states of zero rotational energy. He has also analyzed the case of the passage of a particle through a potential barrier—a case which is, as we know, especially important in the theory of α radioactivity. One quite naturally finds that the trajectories must be very complicated, but the complete indeterminacy assumed by the standard interpretation no longer holds here. It would be the *unknown* initial position of the particle which, by determining its entire subsequent motion, would be the cause of the more or less great rapidity with which it crosses the barrier. The probability which is here involved would then be of the classical type, compatible with causality, arising as a result of our ignorance of the exact initial position of the particle and its subsequent motion.

In studying the case where the Ψ wave is formed by a superposition of stationary waves, Bohm shows by a calculation that the quantum potential, energy and momentum of the particle here fluctuate constantly and very rapidly, and he adds: "If the particle happens to enter a region of space where [the amplitude of the wave] is small, these fluctuations can become quite violent. We see then that, in general, the orbit of a particle in a non-stationary state is very irregular and complicated, resembling Brownian motion more closely than it resembles the smooth track of a planet around the sun." (p. 175, col. 2). We have already noticed the interest this remark holds.

Those are a few of the interesting results developed by Bohm in his two papers, but the most original part of his work is certainly his theory of measurement, which we are now going to analyze.

2. The theory of measurement according to Bohm

Bohm first of all analyzes the phenomena of collision in the causal theory. Before the collision, every particle has its separate packet of Ψ waves in physical space, and the Ψ wave function in configuration space is then simply the product of the individual wave-packets. Then the wave-packets come closer together, overlap, and interaction begins. The Ψ wave function of the system becomes a sum of products, each one of which corresponds to one of the possible results of the interaction. The particle motions then cease to be independent. The a and φ functions, and consequently the quantum potentials and momenta of the particles, undergo rapid and violent fluctuations in space and time. In the regions of configuration space where a is small, the fluctuations are amplified and may result in large exchanges of energy and momenta in a very brief period of time, even when the interaction potential in the usual sense remains small. At the end of the interaction, the wave-packets corresponding to the various possibilities are separated and will no longer overlap in space. Then the representative point of the system will be in one of the wave-packets in configuration space and, for the description in physical space, each particle will have its position in its appropriate wave-packet, and the probability of these correlated positions will be given by $|\Psi|^2$. Since the particles still have a position in physical space, all the difficulties of the current interpretation with regard to correlated systems are removed.

Causality will also be re-established, for the final outcome will depend on the initial positions of the particles; but in practical terms the trajectories of these particles will be extremely complicated and will vary rapidly with the initial positions, and their exact calculation will be impossible. Statistics will thus be brought in, but only in the classical manner; the introduction of statistics will here be necessitated both by the impossibility of exactly determining initial positions without disturbing the whole phenomenon, and by our inability to follow these trajectories; whereas, in the usual interpretation it is assumed that *nothing* determines the result of the interaction, and it is thus taken for granted *a priori* that the description of the phenomena is intrinsically and inevitably statistical.

This study of interaction phenomena led Bohm to analyze measuring processes which, by and large, amount to the interaction of a particle and a measuring apparatus (pp. 179–184 of the second paper). The performance of the measurement establishes "correlations" between the particle's final state and the final state of the measuring apparatus, so that the observation of the final state of the apparatus permits us to deduce the final state of the particle.

Bohm then considers the measurement of a certain "observable" quantity associated with an electron. **R** will represent the electron's position, y the "significant" apparatus coördinate—for example, one that may be read off a graduated dial. Bohm says that it is sufficient to consider an "impulsive" measurement, *i.e.* one involving a very strong and very rapid interaction between the electron and the apparatus. Then the system will not have time to change independently during the interaction, since its entire modification during the measurement will result from that interaction. One can in this way neglect the parts of the Hamiltonian associated with the particle alone or with the apparatus alone, and it will suffice to consider only the Hamiltonian H_1 representing the interaction. Moreover, if H_1 is exclusively a function of quantities that commute with a quantity Q, the interaction process will produce no uncontrollable changes in the quantity Q, but will produce such changes for those which do not commute with Q. In order that the apparatus and the particle shall be coupled, it is necessary that H_1 contain an operator acting upon y.

By way of an example, Bohm puts

$$H_1 = aQ(p_y)_{op} = i\hbar aQ\frac{\partial}{\partial y} \tag{1}$$

where a is a constant.

The development of the Ψ of the system must then be calculated in configuration space **R**, y. During the interaction the wave equation is approximately

$$-i\hbar\frac{\partial\Psi}{\partial t} = H_1\Psi = ia\hbar Q\frac{\partial\Psi}{\partial y} \tag{2}$$

or

$$\frac{\partial\Psi}{\partial t} = -aQ\frac{\partial\Psi}{\partial y}. \tag{3}$$

It is convenient to expand in terms of the complete set of eigenfunctions $\Psi_q(\mathbf{R})$ of the operator Q, where q is an eigenvalue of Q. We then write

$$\Psi(\mathbf{R}, y, t) = \sum_q f_q(y, t)\Psi_q(\mathbf{R}). \tag{4}$$

Since by definition $Q\Psi_q(\mathbf{R}) = q\Psi_q(\mathbf{R})$, the wave equation gives us

$$\frac{\partial}{\partial t}f_q(y, t) = -aq\frac{\partial}{\partial y}f_q(y, t) \tag{5}$$

—an equation whose solution is

$$f_q(y, t) = f_q(y - aqt) \tag{6}$$

whence

$$\Psi(\mathbf{R}, y, t) = \sum_q f_q(y - aqt)\Psi_q(\mathbf{R}). \tag{7}$$

Initially the particle and the apparatus are independent, and one can write for $t = 0$:

$$\Psi_0(\mathbf{R}, y) = g_0(y) \sum_q c_q \Psi_q(\mathbf{R}) \tag{8}$$

because the initial wave function is the product of the initial wave function of the apparatus $g_0(y)$ and the initial wave function of the particle, which can be expanded in terms of the $\Psi_q(\mathbf{R})$ with appropriate coefficients c_q. The wave function $g_0(y)$ will take the form of a packet, which will be assumed centered on the value $y = 0$ with a width Δy.

From the preceding equations it follows that $f_q(y) = c_q g_0(y)$ and that consequently

$$\Psi(\mathbf{R}, y, t) = \sum_q c_q g_0(y - aqt)\Psi_q(\mathbf{R}). \tag{9}$$

This expression shows that the interaction establishes a correlation between q and the apparatus coördinate y.

By then utilizing the results of his analysis of collisions, Bohm notes that, if we write

$$\Psi(\mathbf{R}, y, t) = a(\mathbf{R}, y, t)e^{\frac{1}{\hbar}\phi(\mathbf{R}, y, t)}$$

the real functions a and φ vary extremely rapidly in space and time during the interaction and that the same is true of the quantum potential

$$Q = -\frac{\hbar^2}{2m} \frac{\Delta_{\mathbf{R}} a + \dfrac{\partial^2 a}{\partial y^2}}{a}. \tag{10}$$

From this there results, according to the guidance formula, a very complicated and essentially variable motion of the particle. But, by the end of the interaction *the wave-packets corresponding to the different values of q in $g_0(y - aqt)$ become separated from each other in space.*

In fact, the center of the q wave-packet in the space y is given by

$$y = aqt \qquad \text{or } q = \frac{y}{at} \tag{11}$$

and if we designate by δq the separation of the adjacent eigenvalues of q, we obtain by separation of the centers of the corresponding wave-packets

$$\delta y = at\delta q. \tag{12}$$

For sufficiently large values of t, δy will always become greater than the width Δy of the packets, so that the wave-packets are separated in space.

Bohm concludes from this that, by virtue of the law of probability given by $|\Psi|^2$, the variable y of the apparatus must finally occur in one of the wave-packets and that it will then remain there, since it can no longer take on a value corresponding to the intermediate space between the packets where the probability of presence is zero. If a macroscopic observation then permits us to know the value of q, we will know which probability packet corresponds to the final state actually achieved and, if the function is renormalized to unity, we can say that there corresponds to the final state the wave function

$$\Psi(\mathbf{R}, y, t) = \Psi_q(\mathbf{R})g_0(y - aqt) \tag{13}$$

with the measured value of q.

As one can see in the particular example developed by Bohm, what permits us to explain the possibility of measurement, as well as, in a more general way, to represent the physical separation of the various final states at the end of an interaction, is the limitation of the wave-trains. The limiting of the beams thus plays a very important role here, and, in the appendix of his second Paper, in agreement with what I said at the 1927 Solvay Congress, Bohm considers that this limiting makes it possible to answer the objections which Pauli at that time brought against my theory.

Leaving aside certain thermodynamical considerations (p. 182) of Bohm's Paper which do not seem very clear to us and which might require some adjustment, let us summarize the way in which he brings in probabilities and uncertainty relations. For him, in accord with the ideas of the causal theory, the motions of both the particle and the measuring apparatus are determined by the initial form of the wave $\Psi_0(\mathbf{R}, y)$ and by the *initial values* $\mathbf{R}_0$ and y_0 of the variables that fix the position of the particle and of the apparatus. But since we do not know what these initial values are, and since, moreover, the violent and complicated motion of the system during the interaction is very greatly affected by the slightest variation in the initial conditions, we know only, after the interaction, a probability distribution for the final state given by $|\Psi(\mathbf{R}, y)|^2$. To obtain the probability of a certain value q of Q, we must integrate $|\Psi|^2$ over $\mathbf{R}$ and over y in the q wave-packet of the final state. Since the wave-packets are then separated in configuration space (and likewise in physical space), and since the functions $\Psi_q(\mathbf{R})$ and $g_0(y)$ are normalized, we obtain for this probability the value

$$P_q = |c_q|^2 \tag{14}$$

in accordance with the usual interpretation. More generally, one is able to justify in this way all the rules of the usual interpretation, but by giving them a different meaning.

Let us now suppose that we have obtained, as the result of a measurement, for example, a particle state $\Psi_q(\mathbf{R})$ corresponding to the exact value q of the quantity Q. If this particle interacts with an apparatus designed to measure a quantity P that does not commute with Q, we have for the particle-plus-apparatus system, after the interaction, a wave function of the form

$$\Psi(\mathbf{R}, z, t) = \sum_p a_{pq} \Phi_p(\mathbf{R}) g_0(z - apt) \tag{15}$$

where $\Phi_p(\mathbf{R})$ is the eigen-function of quantity P corresponding to the eigen-value p and where the a_{pq} are the coefficients of the expansion

$$\Psi_q(\mathbf{R}) = \sum_p a_{pq} \Phi_p(\mathbf{R}) \tag{16}$$

and z is the variable of the new measuring apparatus. When the interaction is completed, one will be able, by recording an indication made by the apparatus, to reduce the Ψ function to the single term

$$\Psi = a_{pq} \Phi_p(\mathbf{R}) g_0(z - apt) \tag{17}$$

corresponding to the exact value p of P, and the probability of this happening will be equal to $|a_{pq}|^2$, again completely in accord with the usual interpretation.

Since one and the same apparatus, at the end of its interaction with the particle, will not have been able to "separate" the wave-packets corresponding to the precise values p and q of the non-commuting quantities P and Q, it is obvious that it is impossible to measure such quantities simultaneously, and this permits us to arrive once more, but in a new way, at Heisenberg's uncertainty principle, which here represents not a real indeterminacy of quantities P and Q prior to measurement, but merely the material impossibility of simultaneously obtaining precise values for the two quantites by means of a single measuring operation.

From the point of view adopted here, every quantity Q of the particle has a well defined value in its initial state, but this value is a "hidden variable", since, generally, every attempt at measuring it will result in its modification. If by some exception a measuring apparatus does allow us to obtain a value of Q without modification, then that apparatus will modify the values of all the quantities P that do not commute with Q.

So we must carefully distinguish between the "hidden variables"—which in the causal theory as well as in Classical Physics, would at every instant characterize the particle's position and motion—and the "observables" in Dirac's sense, which are the values of these quantities obtainable by a measuring operation. This shows, in accordance with certain ideas of Bohr but in a totally different manner, the importance of measuring operations. As Bohm quite rightly says, "This means that the measurement of an 'observable' is not really a measurement of any physical property belonging to the observed system alone. Instead, the value of an 'observable' measures only an incompletely predictable and controllable potentiality belonging just as much to the measuring apparatus as to the observed system." (p. 183, vol. 2).

There is, however, an important point on which Bohm, in my opinion, has not sufficiently insisted: That is the fact that in the conceptions of the causal theory, the probability of presence $|\Psi|^2$ has a very special role in connection with the probabilities $|c_p|^2$ of the quantities that do not commute with position. The probability of presence $|\Psi|^2$ is indeed derived from the guidance formula when one does not know (and such is necessarily the case here) which of the trajectories is in fact described, whereas the probabilities $|c_p|^2$ for the values of the quantities that do not commute with position do not enter the picture until after the process of measuring these quantities, when the result of the measurement is not yet known. That is what, from the standpoint of the causal theory, may cast doubt on the absolute correctness of the elegant, but perhaps factitious, "theory of transformations" which seeks to give equal footing to all the expansions of Ψ in keeping with the eigen-functions of all the observables. It is this circumstance that, from the same point of view, may impugn von Neumann's argument concerning hidden variables—as we have previously explained. And this leads us to say a few words about the statistical schema valid in the causal theory.

3. The statistical schema of the causal theory

We already reviewed in an earlier section (Chap. VI, sec. 6) the statistical schema currently used by statisticians. Considering the case of continuous probabilities, one defines for two random quantities X and Y the probability densities $\rho_X(x)$ and $\rho_Y(y)$, as well as the density $\rho(x, y)$ of the probability of simultaneously finding the value x for X and y for Y. The probability of X conditioned by Y is also defined by the density $\rho_X^{(Y)}(x, y)$ of obtaining the value x for X when we know that y has a value Y and the probability of Y conditioned by X by the density $\rho_Y^{(X)}(x, y)$, having a similar definition.

Between these quantities we must have the relations

$$\rho_X(x) = \int dy\, \rho(x, y); \qquad \rho_Y(y) = \int dx\, \rho(x, y)$$
$$\rho_X^{(Y)}(x, y) = \frac{\rho(x, y)}{\rho_Y(y)}; \qquad \rho_Y^{(X)}(x, y) = \frac{\rho(x, y)}{\rho_X(x)} \tag{18}$$

resulting in

$$\rho_X(x) = \int dy\, \rho_X^{(Y)}(x, y)\rho_Y(y)$$
$$\rho_Y(y) = \int dx\, \rho_Y^{(X)}(x, y)\rho_X(x). \tag{19}$$

Now, as we have shown, with the distributions that are assumed in the usual interpretation of Wave Mechanics, the foregoing statistical schema is not applicable to two random variables X and Y corresponding to two non-commuting quantities. The reasons for this are: 1) The quantity $\rho(x, y)$ does not exist, since it is impossible to measure X and Y simultaneously, and 2) A measurement of X alters the probability distribution of Y and vice-versa.

In the causal theory, which is of a more classical type, the situation is quite different, and it is possible to end up with the usual statistical schema. In the causal theory, in fact, the probability distributions of the usual interpretation do not all exist simultaneously because at least a portion of them is created by the very act of measuring. And that is why they do not fit into the usual statistical schema. But the causal theory does allow us to define probabilities that exist simultaneously prior to any measurement and which do satisfy the usual statistical schema.

To simplify: let us consider a problem involving only one spatial variable x; the generalization in three variables would be easy. The random variable corresponding to x will be called X, and we will choose for Y the component p_x of momentum—a quantity that does not commute with X. We then have

$$\rho_X(x) = |\Psi(x)|^2; \qquad \rho_{P_x}^{(X)}(x, p_x) = \delta\left(p_x + \frac{\partial\varphi}{\partial x}\right) \tag{20}$$

by virtue of the statistical meaning of $|\Psi|^2$ and of the guidance formula. For $\rho(x, p_x)$ we find according to (18)

$$\rho(x, p_x) = \rho_X(x)\rho_{P_x}^{(X)}(x, p_x) = |\Psi(x)|^2\delta\left(p_x + \frac{\partial\varphi}{\partial x}\right) \tag{21}$$

and so we have

$$\int dp_x \rho(x, p_x) = |\Psi(x)|^2 = \rho_X(x). \tag{22}$$

We also find

$$\rho_{P_x}(p_x) = \int dx\, \rho(x, p_x) = \int dx\, |\Psi(x)|^2\,\delta\left(p_x + \frac{\partial\varphi}{\partial x}\right) = \sum_i |\Psi(x_i)|^2 \tag{23}$$

the x_i being the values of x for which $-\partial\varphi/\partial x$ has the value p_x considered.

Finally

$$\rho_X^{(P_x)}(x, p_x) = \frac{\rho(x, p_x)}{\rho_{P_x}(p_x)} = \frac{|\Psi(x)|^2\,\delta\left(p_x + \frac{\partial\varphi}{\partial x}\right)}{\sum_i |\Psi(x_i)|^2}. \tag{24}$$

From the collection of all the formulas (20) through (24), we see that it is possible in the causal interpretation to completely restore the usual statistical schema.

4. Takabayasi's observation about the moments of the probability distributions

It is well known that, given a probability distribution defined by $\rho(x)$ for the random variable X, the name "moment of the nth order" of that distribution is given to the quantity

$$M_x^{(n)} = \int dx\, \rho(x)x^n. \tag{25}$$

In his paper of 1952 in *Progress of Theoretical Physics* [5], Takabayasi has calculated and compared the moments that correspond to the probability distributions in the usual interpretation and in the causal interpretation. Let us summarize his results, still confining ourselves to a single variable x.

First of all, with regard to the spatial variable x, it is evident that we have for any positive integer

$$\mathrm{M}_x^{(n)} = \int_{-\infty}^{\infty} dx\, x^n |\Psi(x)|^2 \tag{26}$$

in both the usual and the causal interpretation. More generally, one even has

$$\mathrm{M}_{f(x)}^{(n)} = \int_{-\infty}^{\infty} dx [f(x)]^n |\Psi(x)|^2 \tag{27}$$

in both theories.

But if we consider a quantity that does not commute with X (for example, P_x), one must no longer expect to find all the moments equal in the two theories, for if all the moments were the same, the probability distributions for P_x would be the same in the two theories, and we know that is not so. Indeed, it is shown in the Calculus of Probabilities that the ensemble of the moments completely determines the distribution of the probabilities.

Now if we put

$$\Psi(x) = \int dp_x\, c(p_x) e^{-\frac{i}{\hbar} p_x x}$$

we have

$$\mathrm{M}_{p_x}^{(n)} = \int dp_x\, p_x^n |c(p_x)|^2. \tag{28}$$

In the causal theory, using the expression found above for $\rho(p_x)$, we obtain

$$\mathrm{M}_{p_x}^{(n)} = \int dx\, dp_x\, p_x^n |\Psi(x)|^2 \delta\left(p_x + \frac{\partial \varphi}{\partial x}\right) = \int dx \left(-\frac{\partial \varphi}{\partial x}\right)^n |\Psi(x)|^2. \tag{29}$$

For $n = 1$, we have

$$\mathrm{M}_{p_x}^{(1)} = \int dx \left(i\hbar \frac{\Psi^* \dfrac{\partial \Psi}{\partial x}}{|\Psi|^2}\right) |\Psi|^2 = i\hbar \int dx\, \Psi^* \frac{\partial \Psi}{\partial x}. \tag{30}$$

Replacing Ψ and Ψ^* by their Fourier expansions and taking into account the formula $\delta(p'_x - p_x) = \int dx\, e^{\frac{i}{\hbar}(p'_x - p_x)x}$, we easily arrive at

$$\mathrm{M}_{p_x}^{(1)} = \int\int dp'_x\, dp_x\, p_x c^*(p'_x) c(p_x) \delta(p'_x - p_x) = \int dp_x\, p_x |c(p_x)|^2. \tag{31}$$

The first-order moment of p_x thus has the same value in the two theories.

But for the second-order moment in the causal theory,

$$M_{p_x}^{(2)} = \int dx \left(\frac{\partial \varphi}{\partial x}\right)^2 |\Psi(x)|^2 \tag{32}$$

one can no longer make the analogous calculation, and that moment is not equal to the corresponding moment $\int dp_x \, p_x^2 |c(p_x)|^2$ of the usual theory.

But this does not in any way constitute an objection to the causal theory, because for it the probabilities considered by the usual theory exist only after measurement and are not equal to those existing prior to the measurement. These latter probabilities would be valid for a "super-observer" capable of knowing the value of the quantities directly, without taking any measurements.

It might be thought that the divergence between the results obtained by calculating $M_{p_x}^{(2)}$ in the usual theory and in the causal theory could lead to a decisive choice between the two theories. Would it not suffice to determine $M_{p_x}^{(2)}$ experimentally—which is possible—and see which of the two expressions conforms to reality? Obviously not. Since the experimental determining of $M_{p_x}^{(2)}$ requires that statistical measurements of the quantity P_x be made, the value of $M_{p_x}^{(2)}$ indicated by the usual theory will continue to be the one that represents the observations; but that will prove nothing against the causal theory, since the value of $M_{p_x}^{(2)}$ calculated by it must, by the very nature of the theory, correspond to the unobservable probability distribution existing prior to any measurement.

5. Examination of an observation by Bohr concerning the collision of a particle with an atom

As a conclusion to this chapter, we are going to examine an argument, developed quite a long time ago by Bohr, tending to prove that the ionization of an atom by the collision of a particle cannot be understood by classical ideas on the localization of particles.

We know that if an atomic system is bombarded with high-speed particles, this may set up an excitation and even the ionization of the system. Now this phenomenon, when analyzed according to classical ideas, seems to be incomprehensible. In fact, the incident particle traverses the atom with a velocity v, and if d designates the average diameter of the atom, the transit time of the particle will be of the order d/v. It is solely during a time of this order that the incident particle can act on the atom's components and transfer energy to them in such a way as to cause excitation or ionization. Now in order for a particle of the atomic system to absorb energy, it must be able to change its position considerably during the time $\tau = d/v$. This requires that the time τ shall be at least of the order of the periods T of the motion of the electrons in the atom—periods which are themselves of the order of magnitude of the d divided by the velocity of the atomic electrons. If the incident particle is moving very rapidly, the atomic electrons' velocity will be less than v, and there will be no possible excitation or ionization. More precise calculations confirm this conclusion and show that, with classical conceptions of the localization of particles, the phenomena of excitation and ionization by collision, which are experimentally observed, are inexplicable.

But, says Bohr, such is not the case in the conceptions of the indeterministic theory. In fact, in order to be able to apply the idea of the conservation of energy to the collision, the initial energy of the incident particle must be known to within an uncertainty δE, much smaller than the quantum of energy h/T corresponding to the frequency $\nu = 1/T$ of the motion of the atomic electrons. If that were not the case, the uncertainty δE would be of the order of the energy-differences ΔE between the stationary states of the atom, and it would no longer be possible to verify the conservation of energy. But the wave-train associated with the incident particle always has a finite length and, in keeping with the uncertainty relation $\delta E \delta t \geqq h$, it takes a time δt of the order $h/\delta E$ to pass through the atom. Since in the indeterministic interpretation, in which the incident particle is potentially present throughout the extent of the wave-train, there is no sense in wanting to specify, within the interval δt of time, the instant when the particle enters the atom, it is quite impossible to attribute a value less than δt to the transit-time τ. So we have $\tau \sim h/\delta E \gg h/\Delta E$ and, since the difference ΔE of the energy between the stationary states turns out to be of the order h/T, we will have $\tau \gg T$. The necessary condition for excitation or ionization can thus, in the usual interpretation, be considered to be met, because the duration of the interaction between the atomic components and the incident particle, which is potentially present throughout its wave-train, cannot be regarded as shorter than the total duration of the wave-train's passage through the atom.

It might be thought that this argument of Bohr constitutes an objection to the causal theory. And indeed, in the causal theory, a position, velocity and trajectory are restored to the particles. So an incident particle associated with an approximately monochromatic wave-train would have a well-determined initial velocity v, and one might think that it must traverse the atom in a time of the order d/v; and that would make us vulnerable to Bohr's argument.

In reality, however, it does not seem that Bohr's argument actually does militate against the causal theory. In fact, when the incident wave-train reaches the atom, an interaction begins whose evolution may be represented in terms of the formalism of configuration space. During this interaction the motion of the particle defined by the guidance law will be a very complicated motion associated with the violent and continual variations of the quantum potential. The result is a kind of "buffeting" of the incident particle in the region of the atom which no longer allows us to consider it as passing through this region with a uniform rectilinear motion of velocity v. A possible effect may be that the particle remains in the atomic region for a time that may be of the order τ of the total duration of the wave-train's passage through the atom. Bohr's argument, valid when brought against the application of Classical Dynamics to the problem under consideration, no longer seems to apply here, and the existence of excitations and ionizations by collision does not seem irreconcilable with the causal theory.

Chapter XVI

THE EXTENSION OF THE IDEAS OF THE DOUBLE SOLUTION TO DIRAC'S THEORY OF THE ELECTRON

1. Introduction

We have developed the theory of the Double Solution on the basis of the relativistic wave equation involving a single Ψ (the Klein-Gordon equation) which at the present time must be considered valid only for particles of zero spin. Obviously, we must find forms of the theory that will be applicable to particles of spin higher than zero and especially to particles of spin $\frac{1}{2}$ (in units of $\hbar$), such as the electron.

This extension has been accomplished by Vigier in a form that strikes me as more satisfactory than a different form that I had developed at an earlier date [10]. So it is Vigier's theory that I will present.

2. A summary of Dirac's theory of the electron

First let us recall very summarily the principles of Dirac's theory of the electron (the theory of particles of spin $\frac{1}{2}$). For a more detailed account one may refer to other works on the subject.[1]

In Dirac's theory the Ψ wave of the electron is considered as a four-component quantity Ψ_k (with $k = 1, 2, 3, 4$) and, in order to be able to form linear combinations of Ψ_k, four matrices of four rows and four columns each are introduced. We will adopt for these matrices the form given to them recently by von Neumann,[2] since that form is adapted to relativistic formalism. These matrices γ_i are Hermitian, that is, $(\gamma_i)_{rs} = (\gamma_i)^*_{sr}$. In addition, they satisfy the relations

$$\gamma_i^2 = 1 \qquad \gamma_i\gamma_j + \gamma_j\gamma_i = 0 \qquad \text{for } i \neq j \tag{1}$$

which may be condensed to the form

$$\gamma_i\gamma_j + \gamma_j\gamma_i = 2\delta_{ij}. \tag{2}$$

[1] For example, Louis de Broglie, *La Théorie des particules de spin ½ (électrons de Dirac)*, Gauthier-Villars, Paris, 1952.

[2] *Loc. cit.* p. 68.

The notation $\gamma_i \Psi_k$ then represents, by definition, the linear combination $\sum_{l=1}^{4} (\gamma_i)_{kl} \Psi_l$, and the notation $\Psi_k^* \gamma_i$ in a like manner represents the linear combination $\sum_{l=1}^{4} \Psi_l^* (\gamma_i)_{lk}$ whence $\Psi_k^* \gamma_i = (\gamma_i \Psi_k)^*$ because of the Hermitian nature of the γ_i.

Let us then consider a particle with spin $\frac{1}{2}$ and charge e moving in an electromagnetic field derivable from the potentials **A** and φ. For reasons I will not mention here, Dirac assumed that the four Ψ_k must obey the following wave equations

$$\sum_{i=1}^{4} \gamma_i \left(- i\hbar \frac{\partial}{\partial x_i} + \frac{e}{c} \mathrm{A}_i \right) \Psi_k = - i m_0 c \Psi_k \qquad (k = 1,\ 2, 3, 4) \quad (3)$$

m_0 being the rest-mass of the particle. The x_i are here the coördinates of the position four-vector:

$$x_1 = x, \qquad x_2 = y, \qquad x_3 = z, \qquad x_4 = ict.$$

The A_i are the components of the potential four-vector:

$$\mathrm{A}_1 = \mathrm{A}_x, \qquad \mathrm{A}_2 = \mathrm{A}_y, \qquad \mathrm{A}_3 = \mathrm{A}_z, \qquad \mathrm{A}_4 = i\mathrm{V}.$$

Equations (3) form a system of four simultaneous linear equations with partial derivatives of the first order. By introducing the convention for the summation of indices, which is usual in Relativity, and by suppressing the index k of the Ψ_k, (3) may be written in the abbreviated form

$$\gamma_i \left(- i\hbar \frac{\partial}{\partial x_i} + \frac{e}{c} \mathrm{A}_i \right) \Psi = - i m_0 c \Psi. \qquad (4)$$

Let us introduce the Ψ_k^+ defined by $\Psi_k^+ = \Psi_k^* \gamma_4$ or symbolically $\Psi^+ = \Psi^* \gamma_4$. By using the complex conjugate equation of (3), one obtains as wave equations for the Ψ^+ in the symbolical form analogous to (4)

$$\left(i\hbar \frac{\partial}{\partial x_i} + \frac{e}{c} \mathrm{A}_i \right) \Psi^+ \gamma_i = - i m_0 c \Psi^+. \qquad (5)$$

Let it suffice to note the following points:

1) The four matrices γ_i are not entirely determined by the conditions imposed upon them, but this indeterminacy of the γ_i does not affect the physical predictions that may be made from the theory.

2) If one changes the Galilean reference system by means of a Lorentz transformation, the new wave equations may be written in the same form as the former ones *with the same matrices* γ_i, but the new Ψ_k are linear combinations of the former ones.

3) It can be shown that equations (4) contain the spin of the electron both in its mechanical aspect (proper kinetic moment) and in its electromagnetic aspect (proper magnetic and electric moment).

The densities of the proper magnetic moment and the proper electrical moment are represented in a Galilean system by two space-vectors **M** and **P**. From the relativistic viewpoint these two vectors combine to form a world tensor $\mathcal{M}_{ij}$ which is antisymmetrical of second rank. Its components are

$$\begin{array}{lll} & \mathcal{M}_{14} = -\mathcal{M}_{41} = icP_x & \mathcal{M}_{24} = -\mathcal{M}_{42} = icP_y \\ \mathcal{M}_{ii} = 0 & \mathcal{M}_{34} = -\mathcal{M}_{43} = icP_z & \mathcal{M}_{23} = -\mathcal{M}_{32} = M_x \\ & \mathcal{M}_{31} = -\mathcal{M}_{31} = M_y & \mathcal{M}_{12} = -\mathcal{M}_{21} = M_z. \end{array} \tag{6}$$

From equations (4) and (5) we obtain by multiplying the first one by Ψ^+ on the left and the second by Ψ on the right and by cancelling

$$\Psi^+\gamma_i\frac{\partial}{\partial x_i}\Psi + \frac{\partial \Psi^+}{\partial x_i}\gamma_i\Psi = 0 \tag{7}$$

or

$$\frac{\partial}{\partial x_i}(\Psi^+\gamma_i\Psi) = 0. \tag{8}$$

In this way we have obtained the four-vector

$$j_i = ic\Psi^+\gamma_i\Psi = ic\sum_{\substack{k=1\\l=1}}^{4}\Psi_k^*(\gamma_4\gamma_i)_{kl}\Psi_l. \tag{9}$$

And that, in Dirac's theory, is the "current-density" four-vector whose first three components $\mathbf{j} = ic\Psi^*\gamma\Psi = -c\Psi^+\alpha\Psi$ give us the components of the particles's current density along the space axes, while the fourth $j_4 = ic\Psi^+\gamma_4\Psi = ic\Psi^*\Psi$ represents (within the factor ic introduced by the four-vector notation) the density of the probability of presence

$$\rho = \sum_{k=1}^{4}\Psi_k^*\Psi_k = \sum_{k=1}^{4}|\Psi_k|^2.$$

Note that we obtain the "electrical current-density" four-vector by multiplying the j_i quantities by e.

The four-vector j_i which has just been defined, can be broken down into two parts. To demonstrate this, let us multiply (4) on the left by $\Psi^+\gamma_l$ and (5) on the right by $\gamma_l\Psi$, then add and divide by two. There results, taking (2) into account

$$j_l = i\frac{\hbar}{2m_0}\left(\Psi^+\frac{\partial \Psi}{\partial x_l} - \frac{\partial \Psi^+}{\partial x_l}\Psi\right) - \frac{e}{m_0 c}A_l \Psi^+\Psi - i\frac{\hbar}{2m_0}\sum_{i\neq l}\frac{\partial}{\partial x_i}(\Psi^+\gamma_i\gamma_l\Psi). \quad (10)$$

We can thus break up the four-vector j_l into two four-vectors $j_l^{(1)}$ and $j_l^{(2)}$ defined by

$$\begin{aligned} j_l^{(1)} &= i\frac{\hbar}{2m_0}\left(\Psi^+\frac{\partial \Psi}{\partial x_l} - \frac{\partial \Psi^+}{\partial x_l}\Psi\right) - \frac{e}{m_0 c}A_l \Psi^+\Psi \\ j_l^{(2)} &= -i\frac{\hbar}{2m_0}\sum_{i\neq l}\frac{\partial}{\partial x_i}(\Psi^+\gamma_i\gamma_l\Psi). \end{aligned} \quad (11)$$

This is known as the "Gordon decomposition".

It is easy to find a physical interpretation of $j_l^{(1)}$ and $j_l^{(2)}$. One sees first of all that $j_l^{(1)}$ is the current-density vector corresponding to the overall motion of the particle—for example, by considering the simple case of the plane monochromatic wave in the absence of any field ($A_j = 0$) where one can define a velocity

$$\mathbf{v} = \frac{\mathbf{p}}{m} = \frac{\mathbf{p}}{m_0}\sqrt{1-\beta^2}$$

one finds

$$\frac{1}{ic}j_4^{(1)} = \rho = \Psi^*\Psi \qquad \mathbf{j}^{(1)} = \rho\mathbf{v}.$$

As for the four-vector $j_l^{(2)}$, we interpret it by starting out with an analogy to the classical theory of polarized media. Consider a medium that is the seat of both a magnetic and electrical polarization. Its state of polarization is defined at every point by two space vectors: the vector **M** (density of magnetic moment or intensity of magnetization) and the vector **P** (density of electrical moment or electrical polarization). From the relativistic point of view, these two space vectors combine to form a space-timetensor $\mathcal{M}_{ij}$ anti symmetrical of the second rank whose components are expressed in the notation

$$\mathcal{M}_{ii} = 0 \qquad \begin{aligned} \mathcal{M}_{i4} &= -\mathcal{M}_{4i} = icP_i \\ \mathcal{M}_{ij} &= -\mathcal{M}_{ji} = M_k \end{aligned} \qquad (i = 1, 2, 3) \quad (12)$$

(i, j, k forming an even-numbered permutation of the three indices 1, 2, 3).

Now the classical theory of polarized media shows that this polarization is equivalent to the existence of a microscopic density of charge $\delta = -\operatorname{div} \mathbf{P}$ and to a microscopic current-density $\mathbf{i} = \partial \mathbf{P}/\partial t + \operatorname{curl} \mathbf{M}$. One can then say that the polarization is equivalent to the existence of a current-density vector $\mathbf{j}$ given by

$$j_l = \sum_{i=1}^{4} \frac{\partial}{\partial x_i} \mathcal{M}_{li} \tag{13}$$

because this formula once more gives us the values of δ and $\mathbf{i}$ that we have just found.

If we now compare equation (13) with expression (11) of $j_l^{(2)}$, we see that they may be made identical by putting

$$\mathcal{M}_{ij} = \frac{i\hbar}{2m_0} \Psi^+ \gamma_i \gamma_j \Psi. \tag{14}$$

And that is precisely (within the factor e corresponding to the change from current of probability to the electrical current and thereby causing the appearance of the Bohr magneton $e\hbar/2m_0 c$) the expression for the non-zero components of the antisymmetrical tensor of the second rank which, in Dirac's theory, gives—as we have seen above—the density of the proper magnetic moment and the density of the proper electrical moment, the existence of which is associated with the existence of spin.

Summing up, the current-density four-vector j_l of Dirac's theory breaks down to a four-vector $j_l^{(1)}$ arising from the overall motion (orbital motion) of the particle and to another four-vector $j_l^{(2)}$ arising from the particle's spin.

3. The guidance of the particle in Dirac's theory

For a moment we will adopt the pilot-wave viewpoint and consider the particle as being guided by the Ψ wave. In the Wave Mechanics that utilizes only one wave function, it was sufficient to define the velocity of the particle in terms of the phase φ and the Ψ wave written in the form $\Psi = a e^{\frac{i}{\hbar}\varphi}$. But in the present considerations we run into trouble if we try to transpose this idea. Since there are four components of the Ψ in Dirac's theory, we have to write

$$\Psi_k(x, y, z, t) = a_k(x, y, z, t) e^{\frac{i}{\hbar}\varphi_k(x, y, z, t)} \tag{15}$$

where the functions a_k and φ_k are real. Now there is no reason at all why the φ_k should be all the same; so there is no longer one phase, but four distinct phases. Using as a guide the method that allows us to arrive at the approximation of Geometrical Optics in Dirac's theory, I tried, in a brief paper of September 1952, to define a common phase φ by means of the formula $\varphi_k = \varphi + \varphi'_k$, and since the choice of φ is obviously arbitrary, I sought to determine this choice by means of an auxiliary condition. But since that date I have realized that, since the forms I arrived at led to the definition of the motion by the vector $j_l^{(1)}$ of the Gordon decómposition, this amounted to neglecting the influence of the spin on the motion. Vigier, more correctly, I believe, has sought to define the motion in terms of the overall vector j_l, but in his article on this subject [10] he has nevertheless retained the idea of a common phase φ (which he calls S). I now believe this to be needless and artificial; we should reason directly from the four distinct phases φ_k.

If we start with expression (15) for the Ψ_k, and if we introduce this into the components of the current-density four-vector, expression (10), we obtain

$$j_l = -\frac{1}{m_0}\sum_{k=1}^{4}\left(\frac{\partial\varphi_k}{\partial x_l} + \frac{e}{c}\mathrm{A}_l\right)a_k^+ a_k - i\frac{\hbar}{2m_0}\sum_{i\neq l}\frac{\partial}{\partial x_i}(a_k^+\gamma_i\gamma_l a_k) \quad (16)$$

with

$$a_k^+ = a_k\gamma_4.$$

The particle's motion may be defined by its four-velocity u_l of which the components are

$$u_1 = \frac{dx_1}{ds} = \frac{v_x}{c\sqrt{1-\beta^2}} \qquad u_2 = \frac{dx_2}{ds} = \frac{v_y}{c\sqrt{1-\beta^2}}$$
$$u_3 = \frac{dx_3}{ds} = \frac{v_z}{c\sqrt{1-\beta^2}} \qquad u_4 = \frac{dx_4}{ds} = \frac{i}{\sqrt{1-\beta^2}}. \quad (17)$$

To make the guidance of the particle by the Ψ wave perfectly clear, it seems natural to make the Ψ wave follow one of the lines of current defined by the j_l vector, that is, to put

$$u_i = \mathrm{K}j_i. \quad (18)$$

But the four-vector u_l obeys the relation $\sum_{i=1}^{4} u_i^2 = -1$, from which we conclude

$$\mathrm{K} = \frac{1}{\sqrt{-\sum_{i=1}^{4} j_i^2}}. \quad (19)$$

So Vigier proposes to define the variable proper mass M_0 of the particle by the formula

$$M_0 c u_i = \frac{m_0 j_i}{\Psi^+ \Psi} = \frac{m_0 j_i}{a^+ a} \tag{20}$$

whence

$$M_0 = \frac{m_0}{a^+ a} \frac{j_i}{c u_i} = \frac{m_0}{a^+ a} \frac{1}{c} \sqrt{-\sum_{k=1}^{4} j_k^2}. \tag{21}$$

We will see that this definition [3] of M_0 is nothing less than the generalization within the Dirac theory of the definition that we adopted in relativistic Wave Mechanics utilizing a single Ψ.

To be specific, we introduce the following definitions

$$\overline{\frac{\partial \varphi}{\partial x_j}} = \frac{\sum_{k=1}^{4} \frac{\partial \varphi}{\partial x_j} a_k^+ a_k}{\sum_{k=1}^{4} a_k^+ a_k}$$

and (22)

$$P_j = -\frac{e}{c} A_j + \frac{\frac{\hbar}{i} \sum_{i \neq j} \sum_{k=1}^{4} \frac{\partial}{\partial x_i} (a_k^+ \gamma_i \gamma_j a_k)}{\sum_{k=1}^{4} a_k^+ a_k} = -\frac{e}{c} A_j + \frac{j_j^{(2)}}{a^+ a} m_0.$$

The quantity $(\overline{\partial \varphi / \partial x_j})$ is an average of the four derivatives $\partial \varphi_k / \partial x_j$ when we assign the "weight" $a_k^+ a_k$ to each of them. The four-vector P_j is the sum of the four-vector $-(e/c)A_j$ and a four-vector proportional to $j_j^{(2)}$. With these definitions, one can write, in keeping with (16) and (20)

$$M_0 c u_i = -\overline{\frac{\partial \varphi}{\partial x_i}} + P_i = -\overline{\frac{\partial \varphi}{\partial x_i}} - \frac{e}{c} A_i + P_i' \tag{23}$$

a formula which here plays the role of a guidance formula. The influence of the spin on the motion is represented by the term P_i' proportional to $j_i^{(2)}$.

We wish to show that the preceding formula is actually the generalization of formula (18) of Chapter X,

$$M_0 c u^l = g^{kl} \left(\frac{\partial \varphi}{\partial x^k} - e P_k \right) \tag{24}$$

[3] From this one deduces easily that $M_0^2 = m_0^2 (1 + I_2^2 / I_1^2)$—$I_1$ and I_2 being two classical invariants in Dirac's theory.

which we obtained by starting with the Klein-Gordon equation and which, written in the notation we are now using and utilizing four-vector notation instead of separate space-time coördinates, becomes

$$M_0 c u_i = -\left(\frac{\partial\varphi}{\partial x_i} + \frac{e}{c} A_i\right) = -\frac{\partial\varphi}{\partial x_i} - \frac{e}{c} A_i. \tag{25}$$

To see this, one has only to note that, in order to move from the Dirac to the Klein-Gordon theory, we must, first, neglect the effect of spin, *i.e.* the term P'_i in (23); and second, assume that the Ψ_k may be reduced to a single $\Psi = a e^{\frac{1}{\hbar}\varphi}$, which gives us $(\overline{\partial\varphi/\partial x_i}) = \partial\varphi/\partial x_i$. So the agreement between (23) and (25) is achieved.

We easily obtain from (23) the relations

$$\frac{d}{ds}(M_0 c u_i) = \sum_j \frac{\partial}{\partial x_j}(M_0 c u_i)\frac{dx_j}{ds} = \sum_j u_j \frac{\partial}{\partial x_j}\left(-\frac{\overline{\partial\varphi}}{\partial x_i} + P_i\right) \tag{26}$$

$$= \sum_j \left\{u_j \frac{\partial}{\partial x_i}\left(-\frac{\overline{\partial\varphi}}{\partial x_j} + P_j\right) + u_j\left[\frac{\partial}{\partial x_j}\left(P_i - \frac{\overline{\partial\varphi}}{\partial x_i}\right) - \frac{\partial}{\partial x_i}\left(P_j - \frac{\overline{\partial\varphi}}{\partial x_j}\right)\right]\right\}.$$

Since $\sum_j u_j^2 = -1$, one has $\sum_j u_j\, \partial u_j/\partial x_i = 0$, and the first term of the last member is equivalent to $c(\partial M_0/\partial x_i)$ so that we finally have as the equation of motion

$$\frac{d}{ds}(M_0 c u_i) = c\frac{\partial M_0}{\partial x_i} + \sum_j u_j F_{ij} + \sum_j u_j\left[\frac{\partial}{\partial x_j}\left(P'_i - \frac{\overline{\partial\varphi}}{\partial x_i}\right) - \frac{\partial}{\partial x_i}\left(P'_j - \frac{\overline{\partial\varphi}}{\partial x_j}\right)\right] \tag{27}$$

where $F_{ij} = -(e/c)(\partial A_i/\partial x_j - \partial A_j/\partial x_i)$ is the product of the electromagnetic field and the charge.

The last term of (27) expresses the action of the spin on the motion; it is obviously zero in the case of the absence of spin, since the four-vector P'_i is zero in that case, and since—in view of the fact that $(\overline{\partial\varphi/\partial x_i})$ reduces to $\partial\varphi/\partial x_i$—we have

$$\frac{\partial}{\partial x_j}\left(\frac{\overline{\partial\varphi}}{\partial x_i}\right) = \frac{\partial}{\partial x_i}\left(\frac{\overline{\partial\varphi}}{\partial x_j}\right).$$

If the spin is zero or exerts only a negligible effect, we again have the equation

$$\frac{d}{ds}(M_0 c u_i) = c\frac{\partial M_0}{\partial x_i} + \sum_j u_j F_{ij} \tag{28}$$

clearly corresponding to the Dynamics studied in Chapter X.

In this way it is easy to verify that definition (21) for M_0 reverts, when spin is neglected, to the definition previously adopted when we started with the Klein-Gordon equation. As a matter of fact, if spin is neglected, one has

$$j_i^{(2)} = 0 \qquad \text{and} \qquad j_l = j_l^{(1)}$$

and one may consider a single component of the Ψ, in short, a single a and a single φ. One then finds

$$j_i^{(1)} = -\frac{a^2}{m_0}\left(\frac{\partial\varphi}{\partial x_i} - \frac{e}{c}A_i\right)$$

$$-\sum_{i=1}^{4} j_i^2 = -\sum_{i=1}^{4} j_i^{(1)2} = -\frac{a^4}{m_0^2}\sum_{i=1}^{4}\left(\frac{\partial\varphi}{\partial x_i} - \frac{e}{c}A_i\right)^2. \quad (29)$$

Definition (21) then gives us

$$M_0^2 c^2 = -\sum_{i=1}^{4}\left(\frac{\partial\varphi}{\partial x_i} - \frac{e}{c}A_i\right)^2. \quad (30)$$

Now the generalized Jacobi equation gives us in its relativistic form

$$-\sum_{i=1}^{4}\left(\frac{\partial\varphi}{\partial x_i} - \frac{e}{c}A_i\right)^2$$

$$= \frac{1}{c^2}\left(\frac{\partial\varphi}{\partial t} - eV\right)^2 - \sum_{xyz}\left(\frac{\partial\varphi}{\partial x} + \frac{e}{c}A_x\right)^2 = m_0^2c^2 + \hbar^2\frac{\Box a}{a} \quad (31)$$

which brings us back to the expression for M_0 deduced from the Klein-Gordon equation.

In short, by adopting definition (21) for M_0 and by deducing formula (23) from it, we have obtained a natural generalization of the guidance formula in Dirac's theory. The world line that is thus imposed on the particle is one of the lines of current defined by the four-vector j_l. As previously, but always with the reservation that a demonstration of ergodicity must still be supplied, we conclude that the density of the probability of presence is given by

$$\rho = \frac{j_4}{ic} = \sum_{k=1}^{4}\Psi_k^*\Psi_k = |\Psi|^2. \quad (32)$$

The extension of the idea of guidance that we have just presented, by moving from the Klein-Gordon equations to those of Dirac, should also be realizable for particles of spin greater than $\frac{1}{2}$. Vigier has shown that such is indeed the case in the framework of the formalism introduced by the theory of photons.

4. The introduction of the double-solution idea into Dirac's theory

We have just seen how the idea of the guidance of the particle by the Ψ wave may be introduced into Dirac's theory, and this constitutes the pilot-wave viewpoint. Since that view is certainly inadequate, there is need here for the introduction of the double-solution idea, *i.e.* to assume that for every continuous solution $\Psi_k = a_k e^{\frac{i}{\hbar}\varphi_k}$ of Dirac's equations there must be a corresponding solution

$$u_k = f_k(x, y, z, t) e^{\frac{i}{\hbar}\varphi_k(x, y, z, t)}$$

with the same φ_k, where the f_k involve one mobile singular point, or rather, one small mobile singular region wherein the u_k would have a high value and obey non-linear equations.

In 1927, within the framework of relativistic Wave Mechanics involving only a single Ψ, we showed [4] that, in the case of uniform rectilinear motion in the absence of any field, there existed in the proper system of a particle of rest-mass m_0 a solution involving a singularity and spherical symmetry:

$$u_0(x_0, y_0, z_0, t_0) = \frac{1}{r_0} e^{\frac{i}{\hbar} m_0 c^2 t_0} \tag{33}$$

and one had only to effect a Lorentz transformation to obtain, in another Galilean system, the wave function $u(x, y, z, t)$ having a mobile singularity that would represent the particle in motion. Of course, if one substitutes the idea of a small singular region for that of a point-singularity, one must consider that the solution obtained represents the u wave of a particle having spherical internal symmetry, but represents the u wave *only outside of the singular region.*

It is natural to look for singularity solutions, within Dirac's theory, analogous to the one just given. Let us consider a Dirac electron in the absence of any field, and let us put ourselves in its rest system. The "small components" u_1 and u_2 will be assumed of zero value, and, by

[4] See Chapter IX, section 3.

adopting the usual form of the Dirac matrices, we will find for u_3 and u_4 the equations [5]

$$\left(\frac{\partial}{\partial x}+i\frac{\partial}{\partial y}\right)u_4+\frac{\partial u_3}{\partial z}=0 \qquad \left(\frac{\partial}{\partial x}-i\frac{\partial}{\partial y}\right)u_3-\frac{\partial u_4}{\partial z}=0$$
$$\frac{\hbar}{ic}\frac{\partial u_3}{\partial t}=m_0 c u_3 \qquad \frac{\hbar}{ic}\frac{\partial u_4}{\partial t}=m_0 c u_4. \tag{34}$$

The two latter equations are satisfied by putting

$$u_3 = f_3 e^{\frac{i}{\hbar}m_0 c^2 t}, \qquad u_4 = f_4 e^{\frac{i}{\hbar}m_0 c^2 t} \tag{35}$$

which means that

$$\Delta f_3 = \Delta f_4 = 0.$$

If we introduce the variables

$$u = \tfrac{1}{2}(x+iy) \qquad v = \tfrac{1}{2}(x-iy) \tag{36}$$

we obtain

$$\frac{\partial f_4}{\partial v} = -\frac{\partial f_3}{\partial z}, \qquad \frac{\partial f_3}{\partial u} = +\frac{\partial f_4}{\partial z}. \tag{37}$$

These equations give the singularity solution

$$f_4 = \frac{\partial^2}{\partial u \partial z}\left(\frac{1}{r}\right) \qquad f_3 = \frac{\partial^2}{\partial z^2}\left(\frac{1}{r}\right) \tag{38}$$

as we may immediately verify by recalling that $\Delta(1/r) = 0$.

Returning to the variables x and y, we note that $r = \sqrt{x^2+y^2+z^2}$ is equal to $\sqrt{4uv+z^2}$. And we see that, in the proper system, a singularity solution of the type $u_k = f_k e^{\frac{i}{\hbar}m_0 c^2 t}$ has been obtained, with

$$f_1 = f_2 = 0 \qquad f_3 = \frac{2z^2 - x^2 - y^2}{r^5}; \qquad f_4 = \frac{3(x-iy)z}{r^5}. \tag{39}$$

By effecting a Lorentz transformation, and by taking into account the corresponding transformation of the Dirac components, one would obtain the form of u_k in another Galilean system where the singularity would have a uniform rectilinear motion.

Since equations (37) remain intact when simultaneously, f_3 is per-

[5] In order to simplify the notation, we will omit the zero subscript on the variables x, y, z, t in the electron's rest system.

muted with f_4 and, u with $-v$, one deduces the existence of another singularity solution given in the proper system by

$$f_1 = f_2 = 0, \qquad f_3 = -\frac{\partial^2}{\partial v \partial z}\left(\frac{1}{r}\right) = -\frac{3(x+iy)}{r^5}z$$
$$f_4 = \frac{\partial^2}{\partial z^2}\left(\frac{1}{r}\right) = \frac{2z^2 - x^2 - y^2}{r^5}. \tag{40}$$

It will be noted that solutions (39) and (40) do not possess spherical symmetry; they correspond, not to isolated poles, but to dipoles. For particles of spin greater than $\frac{1}{2}$, one would find singularity solutions corresponding to multipoles of a higher order.[6]

If for the notion of point-singularity one substitutes the idea of a small singular region where the equation for u is no longer linear, one must consider the solutions indicated above as being valid only outside the small singular region.

Let us now examine how the question of the relationship between the u and Ψ waves appears in the Dirac theory.

To see this more clearly, let us first take up the question in the theory that omits spin. Equation (23), when one neglects the term P'_i arising from the spin, and when one considers a single phase φ, is written

$$M_0 c u_i = -\frac{\partial \varphi}{\partial x_i} - \frac{e}{c}A_i \qquad (i = 1, 2, 3, 4) \tag{41}$$

which gives

$$\frac{M_0 v_i}{\sqrt{1-\beta^2}} = -\frac{\partial \varphi}{\partial x_i} - \frac{e}{c}A_i \qquad (i = 1, 2, 3)$$
$$\frac{M_0 c_i}{\sqrt{1-\beta^2}} = -\frac{1}{ic}\frac{\partial \varphi}{\partial t} - \frac{e}{c}iV \tag{42}$$

whence

$$\frac{M_0}{\sqrt{1-\beta^2}} = \frac{1}{c^2}\left(\frac{\partial \varphi}{\partial t} - eV\right). \tag{43}$$

By division, one then obtains

$$v_i = -c^2 \frac{\dfrac{\partial \varphi}{\partial x_i} + \dfrac{e}{c}A_i}{\dfrac{\partial \varphi}{\partial t} - eV} \tag{44}$$

[6] Gérard Petiau has made a very complete study of these other kinds of solutions. (*Compt. rend.*, 238 (1954) 998.)

which is exactly the same guidance formula previously obtained in the theory without spin [Chapter IX, formula (36)]. The velocity is then entirely determined by the phase φ common to both u and Ψ. The generalized Jacobi equation (J) valid for u and Ψ then imposes, as we have seen, the equality of the ratios $\Box a/a$ and $\Box f/f$, which permits us to express the mass M_0 in two equivalent forms

$$M_0 = \sqrt{m_0^2 + \frac{\hbar^2}{c^2}\left(\frac{\Box a}{a}\right)} = \sqrt{m_0^2 + \frac{\hbar^2}{c^2}\left(\frac{\Box f}{f}\right)}. \tag{45}$$

Let us now see what happens in Dirac's theory. Formula (23) gives us

$$\frac{M_0 v_i}{\sqrt{1-\beta^2}} = -\frac{\overline{\partial\varphi}}{\partial x_i} - \frac{e}{c}A_i + P'_i \quad (i = 1, 2, 3)$$
$$\frac{M_0 ci}{\sqrt{1-\beta^2}} = -\frac{1}{ic}\frac{\overline{\partial\varphi}}{\partial t} - \frac{e}{c}iV + P'_4 \tag{46}$$

whence

$$\frac{M_0}{\sqrt{1-\beta^2}} = \frac{1}{c^2}\left(\frac{\overline{\partial\varphi}}{\partial t} - eV\right) + \frac{P'_4}{ic} \tag{47}$$

with

$$\frac{P'_4}{ic} = \frac{\rho^{(2)} m_0}{a^+ a} = \frac{\frac{\hbar}{i}\sum_{i=1}^{4}\frac{\partial}{\partial x_i}(a^+\gamma_i\gamma_4 a)}{a^+ a}. \tag{48}$$

P'_4/ic is a real quantity. Finally, by division we obtain from the two formulas (46)

$$v_i = -c^2\frac{\frac{\overline{\partial\varphi}}{\partial x_i} + \frac{e}{c}A_i + P'_i}{\frac{\overline{\partial\varphi}}{\partial t} - eV + \frac{P'_4}{ic}} \quad (i = 1, 2, 3). \tag{49}$$

And that is the exact form of the guidance formula in Dirac's theory.

5. The consequences of these formulas

The double-solution point of view leads us to assume that the velocity (of three components) and proper-mass M_0 given by formulas (49) and (21) must have the same value whether calculated on the basis of the u wave or the Ψ wave. Now, since the four $\overline{\partial\varphi}/\partial x_i$ and the four P_i here depend on a_k, it is no longer enough to suppose that every

u_k has the same phase φ_k as the corresponding Ψ_k. In addition, when one substitutes the f_k for the a_k, the expressions $-(\overline{\partial\varphi/\partial x_i})-(e/c)A_i+P_i'$ for $i = 1, 2, 3, 4$ must retain the same values, so that the velocity v_i and proper-mass M_0 also retain the same value.

At first sight this condition, which seems indispensable if one is to introduce the Double Solution into Dirac's theory, appears extremely restrictive. We will see, however, that it arises quite naturally from an idea which we here meet for the first time, but which will reappear in the next chapter.

In short, what we are led to postulate is that the lines of current defined by the formulas of Dirac's theory, based either on the u or the Ψ wave, coincide. But of course, they can coincide only in the region of space where the u wave obeys the same linear equation as the Ψ wave does, that is, outside of the very small mobile singular region where u has very high values and no longer obeys the linear equation of Ψ. The statement that the quasi-point-like singular region follows one of the lines of current defined by the four-vector j_i(j_i being calculated in terms of the components u_k of u) can have no meaning unless j_i is calculated along the periphery of the singular region—for example, on the sphere S, which we introduced in Chapter IX, section 6, for the demonstration of the guidance formula and which, since it encloses all of the small singular region, is itself already in the region where u obeys a linear equation of propagation. It is, of course, also necessary that the singular region be so small that the continuous Ψ wave will be approximately constant over the whole of this region, including the sphere S. Then, on the whole sphere S, the space vectors $\mathbf{j}(j_1, j_2, j_3)$ are equal and parallel, and one should be able to calculate them equally well in terms of either u_k or Ψ_k.

At any point on the sphere S, the u wave which at that point obeys Dirac's linear equations satisfies the continuity equation

$$\frac{\partial}{\partial t}|u|^2 + \text{div}\,(|u|^2\mathbf{v}) = 0 \tag{50}$$

where $\mathbf{v}$ is the velocity locally defined in space by means of the current-density vector of Dirac's theory. By noting that

$$|u|^2 = \sum_{k=1}^{4}|u_k|^2 = \sum_{k=1}^{4}|f_k|^2 = |f|^2$$

one can write equation (50) in the form

$$\frac{\partial}{\partial t} f^2 + \mathbf{v} \cdot \text{grad}\, f^2 + f^2 \,\text{div}\, \mathbf{v} = 0 \tag{51}$$

or if we call s the space variable taken along $\mathbf{v}$,

$$\frac{\partial}{\partial t} f^2 + v \frac{\partial f^2}{\partial s} + f^2 \text{ div } \mathbf{v} = 0. \tag{52}$$

Now we have assumed (Chap. IX, sec. 6) that on the sphere S, even though the linear wave equations are approximately valid for u, the function f begins to increase rapidly, so that $(\partial/\partial s)\, f^2 \gg f^2$. One can then neglect the last term of the preceding equation, and one finds

$$v = -\frac{\dfrac{\partial f^2}{\partial s}}{\dfrac{\partial f^2}{\partial t}} = -\frac{\dfrac{\partial f}{\partial s}}{\dfrac{\partial f}{\partial t}}. \tag{53}$$

So we see that at any point on S the values of the amplitude $f = \sqrt{\sum_{k=1}^{4} |f_k|^2}$ of the u wave move in the direction $\mathbf{v}$ with velocity given by (53). This means that the point-like singular region of u is displaced in an overall motion with velocity v along the lines of current defined by the space vectors $\mathbf{j}$ which are by hypothesis all equal and parallel at every point on S. And in this way we have effected the demonstration of the guidance formula within Dirac's theory.

But the foregoing considerations, in obliging us to admit that the calculation of the four-vector j_i on S leads to the same result whether we start with u_k or Ψ_k, almost inevitably leads us to consider the following idea: "Outside the small singular region the function u, which inside that region obeys a non-linear equation, would be proportional to the normalized Ψ function; in other words, outside of S we would have $u \simeq C\Psi$." The constant C would have a well determined value, since u, having an objective significance, must have a well determined value and cannot be arbitrarily normalized the way Ψ can.

In the theory without spin this hypothesis leads us to put

$$f = |C| a, \qquad \text{phase of } u = \text{phase of } \Psi + \text{const.} \tag{54}$$

The second relation expresses the equality of the phases (within a constant—a fact of no importance since guidance here requires the introduction of only the derivatives of φ). The first relation (54) obviously gives $\Box f/f = \Box a/a$ and we have thus found once more the relations postulated by the hypothesis of the Double Solution in spin zero Wave Mechanics.

In Dirac's theory, the hypothesis $u = C\Psi$ gives

$$f_k = |C|a_k, \qquad \text{phase of } u_k = \text{phase of } \Psi_k + \arg C. \tag{55}$$

The second relation (55) expresses the postulate that was introduced when we assumed that the components of the same index u_k and Ψ_k had the same phase φ_k. Moreover, the ensemble of relations (55) shows that the four-vector defined by

$$j_i = ic \sum_{\substack{k=1\\l=1}}^{4} \Psi_k^* (\gamma_4 \gamma_i)_{kl} \Psi_l \quad j_i = ic \sum_{\substack{k=1\\l=1}}^{4} u_k^* (\gamma_4 \gamma_i)_{kl} u_l \tag{56}$$

coincide on S and outside of S within the multiplicative constant $|C|^2$. It is easy to verify that the guidance-velocity $\mathbf{v}$ and the proper-mass M_0 then have the same value, whether calculated on S in terms of u or Ψ.

In the next chapter we are going to examine more closely the new idea that we have been led to introduce, and we will see the full extent of its scope.[7]

[7] The hypothesis expressed by $u = C\Psi$ (outside the singular region) gives rise, in Dirac's theory, to a difficulty that is not present in the case of Wave Mechanics utilizing a single Ψ.

As a matter of fact, in Dirac's theory the choice of the γ (or α) matrices is largely arbitrary, and, depending upon the way one makes this choice, the form of the Ψ_k varies. Only the form of the bilinear quantities in Ψ and Ψ^* remains invariant—quantities which, in the usual interpretation, constitute the only ones in the theory which have a physical meaning. Since the four-vector j_i is precisely a quantity of this type, the guidance defined by j_i is independent of the indeterminacy of form of the Ψ_k. But since, in the theory of the Double Solution, we wish to give the u wave a physical meaning, it seems that one must assume that the components u_k have well determined values. In that case, in the linear equations valid for the u_k outside the singular region, it must be assumed that the γ matrices have a physically determined form, and the relation $u = C\Psi$ will not be possible unless we take, for the Ψ_k, a solution of the Dirac equations written with exactly this form of the γ matrices. The indeterminacy of the γ matrices, and of the Ψ_k components resulting from them, would thus only arise when, by adopting the point of view of the usual interpretation, we omit the u wave and preserve only the continuous Ψ wave, having the character of a fictitious probability wave.

Chapter XVII

THE STRUCTURE OF THE u WAVE AND ITS RELATION TO THE Ψ WAVE

1. The difficulty of proving the existence and of determining the form of the u wave

In order to be able to make a really complete presentation of the theory of the Double Solution, one would have to prove the existence of the u wave, determine its form and show exactly what relations exist between it and the Ψ wave usually considered in Wave Mechanics. At this point a whole series of difficult questions arises, which we will now examine, at least in part. So long as these questions are not clearly answered, there can be no possibility of constructing a truly coherent causal interpretation of Wave Mechanics.

When I wrote my 1927 Paper, I considered the u wave as obeying a linear equation of propagation involving a genuine mathematical singularity. I naturally felt the necessity of demonstrating the existence of these u waves involving a singularity. Since I had found a singularity-solution for the case of the absence of any field, it was my hope that this solution could be generalized by the use of an argument something like the following: In a field-free region R_0 there exist u and Ψ solutions, paired in the way required by the principle of the Double Solution; if these waves then penetrate a region R where there is a field, the singularity of the u wave cannot cease to exist, and one could try to demonstrate that the prolongation of the u wave in the region R continues to be linked with the prolongation of the Ψ wave by virtue of the requirements of the theory of the Double Solution. But I was unable to find any precise form for this demonstration.

Summing up, the theory I tried to develop presupposed the following existence-theorem:

Given a certain equation of propagation valid within a certain region of space, and assuming that within this region a continuous Ψ solution exists, such that

$$\Psi(x, y, z, t) = a(x, y, z, t)e^{\frac{i}{\hbar}\varphi(x,y,z,t)}$$

satisfying certain boundary conditions, then there exists another solution

$$u(x, y, z, t) = f(x, y, z, t)e^{\frac{i}{\hbar}\varphi(x, y, z, t)}$$

having the same phase $\varphi(x, y, z, t)$ *and obeying the same boundary conditions and having an amplitude involving a singularity that is in general mobile.*[1]

In this way the problem is formulated but not solved. Moreover, the existence-theorem could have been stated a little differently. Let us write the two waves in the form

$$\begin{aligned} \Psi(x, y, z, t) &= a(x, y, z, t)e^{\frac{i}{\hbar}\varphi(x, y, z, t)}; \\ u(x, y, z, t) &= f(x, y, z, t)e^{\frac{i}{\hbar}\varphi'(x, y, z, t)}. \end{aligned} \tag{1}$$

In our previous formulation we have assumed the "concordance of phases" in the strict form

$$\varphi(x, y, z, t) = \varphi'(x, y, z, t) \tag{2}$$

[1] It might be well to point out how the problem of the existence of the u wave in the absence of field may be formulated.

In *Euclidean* space-time, the equation $\Box\Psi + m_0^2c^2/\hbar^2\Psi = 0$ is written in curvilinear coördinates $x^\alpha = F_\alpha(x, y, z, t)$ in the form

$$\frac{1}{\sqrt{-g}}\frac{\partial}{\partial x^\alpha}\sqrt{-g}g^{\alpha\beta}\frac{\partial\Psi}{\partial x^\beta} + \frac{m_0^2c^2}{\hbar^2}\Psi = 0. \tag{a}$$

By putting $\Psi = a^{\frac{i}{\hbar}\varphi}$ with a and φ real, we obtain the two equations

$$\text{(J)} \qquad \left[g^{\alpha\beta}\frac{\partial\varphi}{\partial x^\alpha}\frac{\partial\varphi}{\partial x^\beta} - m_0^2c^2\right]a = \hbar^2\frac{1}{\sqrt{-g}}\frac{\partial}{\partial x^\alpha}\sqrt{-g}\,g^{\alpha\beta}\frac{\partial a}{\partial x^\beta}$$

$$\text{(C)} \qquad 2g^{\alpha\beta}\frac{\partial a}{\partial x^\alpha}\frac{\partial\varphi}{\partial x^\beta} + \frac{a}{\sqrt{-g}}\frac{\partial}{\partial x^\alpha}\sqrt{-g}\,g^{\alpha\beta}\frac{\partial\varphi}{\partial x^\beta} = 0.$$

Let us consider the hypersurfaces $\varphi(x, y, z, t) = \text{const.} = x^4$ with "internal" metrics $d\sigma^2 = g_{ik}dx^idx^k$ with $i, k = 1, 2, 3$ and the Γ curves orthogonal in space-time to this family of hypersurfaces. The Γ curves are defined by $u^\alpha = g^{\alpha\beta}\,\partial\varphi/\partial x^\beta$ and coincide with the lines of current determined by the guidance formula. Since the hypersurfaces are not, in general, geodesically parallel, we have for every point in Euclidean space

$$ds^2 = dS_\Gamma^2 + d\sigma^2 = g_{44}(x^1, x^2, x^3, x^4)(dx^4)^2 + g_{ik}(x^1, x^2, x^3, x^4)dx^i dx^k \tag{b}$$

whence

$$g^{44} = \frac{1}{g_{44}} \qquad g^{i4} = 0. \tag{c}$$

Equation (C) shows that since $a^2\sqrt{-g}\,g^{44}$ is independent of x^4 and, since from this for every ensemble of values x, y, z, t. That, naturally, is a requirement that seems very severe. Now this postulate serves chiefly to demonstrate the guidance formula and to effect the transition from single-particle Wave Mechanics to the Wave Mechanics of systems of particles in configuration space. If we observe how hypothesis (2) enters the reasoning here, we note that it is not absolutely necessary to assume the concordance of the phases φ and φ' in the form (2). It would be quite sufficient to assume that the phases φ and φ' coincide in the immediate vicinity of the particle.

For reasons that will presently be explained, we have been led to substitute for the idea that the u wave involves a point singularity, the somewhat different idea of a very small singular region, generally mobile, in which the wave would obey a non-linear wave equation. It would then suffice to postulate that the phases φ and φ', as well as their first-order derivatives, have the same values on the tiny sphere S, surrounding the singular region, which helped us in demonstrating the guidance formula.

---

we conclude that $\partial/\partial x^\alpha(a^2\sqrt{-g}\,u^\alpha) = 0$, this expresses the conservation of the fluid whose proper density is $\varrho_0 = a^2\sqrt{-g}$. As for equation (J), it gives

$$\frac{1}{\hbar^2}\frac{1}{\sqrt{-g}}\frac{\partial}{\partial x^\alpha}\sqrt{-g}\,g^{\alpha\beta}\frac{\partial a}{\partial x^\beta} = (g^{44} - m_0^2c^2)a. \qquad (d)$$

One thus sees that $1/c\sqrt{g^{44}}$ is equal to the variable rest-mass M_0 previously defined.

Equation (d) combined with the boundary conditions determines a Ψ solution that is finite, uniform and continuous and which, for a Galilean observer, will have the form

$$\Psi = a(x^1, x^2, x^3, x^4)e^{\frac{i}{\hbar}x_4} = a(x, y, z, t)e^{\frac{i}{\hbar}F_4(x,y,z,t)}. \qquad (e)$$

Ψ, and consequently φ, being thus determined, let us consider the equation of propagation which is satisfied by the u wave of the Double Solution, outside the singular region:

$$\frac{1}{\sqrt{-g}}\frac{\partial}{\partial x^\alpha}\sqrt{-g}\,g^{\alpha\beta}\frac{\partial u}{\partial x^\beta} + \frac{m_0^2c^2}{\hbar^2}u = 0. \qquad (f)$$

By putting $u = f(x^1, x^2, x^3, x^4)e^{\frac{i}{\hbar}x_4}$, we obtain, by substituting in (f) the same equations (J) and (C) as above except for the replacing of a by f. Equation (C) shows still further that $f^2\sqrt{-g}\,g^{44}$ is independent of x^4 and equation (J) gives

$$\frac{1}{\sqrt{-g}}\frac{\partial}{\partial x^\alpha}\sqrt{-g}\,g^{\alpha\beta}\frac{\partial f}{\partial x^\beta} = (g^{44} - m_0^2c^2)f. \qquad (g)$$

It would have to be demonstrated that solutions of this equation exist which possess a singular space-time line coinciding with one of the Γ curves.

Our chief concern here is simply to point out this possibility of weakening postulate (2) concerning the concordance of phases. This question will be taken up again a little further along. (See Section 9).

2. A theorem concerning the Green's functions of the wave equation

We now consider a difficulty that arose as a result of the original form of my conceptions of the u wave.

The difficulty in question results from a theorem concerning the Green's functions of the wave equation—a theorem which, first pointed out by Lord Rayleigh, has been utilized by Sommerfeld in his works on Wave Mechanics.

Let us consider a quantized system, a hydrogen atom, for example. We know that in an s state of this system the particle, according to the guidance formula, must remain motionless at a point Q. If the u wave involves a point-singularity, that singularity must be found at Q. The equation for the Ψ waves is here of the form

$$\Delta\Psi + (k^2 - F(r))\Psi = 0 \tag{3}$$

where $F(r)$ is the Coulomb potential of the nucleus of the H atom. The eigen values k_i and the corresponding eigen functions Ψ_i are defined by

$$\Delta\Psi_i + [k_i^2 - F(r)]\Psi_i = 0. \tag{4}$$

If we assume that there exists a solution of the wave equation having a singularity like $1/r$ at the point Q and of zero-value at the boundaries of the region [that will be the Green's function of equation (3) relative to the point Q and to the region under consideration], we should have for u

$$\Delta u + [k^2 - F(r)]u = \varepsilon\delta(M - Q) \tag{5}$$

M being the current point and $\delta(M - Q)$ the singular Dirac function relative to the point Q. The wave equation will thus be everywhere satisfied by u, except at the point Q where a singularity like $1/r$ will exist.

But $\delta(M - Q)$ may be expanded by means of the eigen functions $\Psi_i(M)$ in the form

$$\delta(M - Q) = \sum_i c_i\Psi_i(M) \qquad c_i = \int d\tau\, \delta(M - Q)\Psi_i^*(M) = \Psi_i^*(Q) \tag{6}$$

whence

$$\delta(\mathrm{M}-\mathrm{Q}) = \sum_i \Psi_i^*(\mathrm{Q})\Psi_i(\mathrm{M}). \tag{7}$$

If we then expand $u(\mathrm{M})$ in the form

$$u(\mathrm{M}) = \sum_i c_i \Psi_i(\mathrm{M}) \tag{8}$$

with new coefficients c_i, we must have

$$[\Delta + k^2 - \mathrm{F}(r)] \sum_i c_i \Psi_i(\mathrm{M}) = \varepsilon \sum_i \Psi_i^*(\mathrm{Q})\Psi_i(\mathrm{M}) \tag{9}$$

or, according to (4)

$$\sum_i c_i(k^2 - k_i^2)\Psi_i(\mathrm{M}) = \varepsilon \sum_i \Psi_i^*(\mathrm{Q})\Psi_i(\mathrm{M}) \tag{10}$$

from which we obtain, since the Ψ_i form a complete system

$$c_i = \varepsilon \frac{\Psi_i^*(\mathrm{Q})}{k^2 - k_i^2} \tag{11}$$

then

$$u(\mathrm{M}) = \sum_i \frac{\varepsilon \Psi_i^*(\mathrm{Q})\Psi_i(\mathrm{M})}{k^2 - k_i^2}. \tag{12}$$

This expression of the Green's function $u(\mathrm{M}, \mathrm{Q})$ constitutes the theorem I have referred to.

Now in a stationary state the u function must have a frequency equal to the stationary state considered, that is, k must be equal to one of the k_i. We then see that the coefficient of $\Psi_i(\mathrm{M})$ in (12) is infinite except when $\Psi_i(\mathrm{Q}) = 0$. In other words, the singularity is situated at a point where Ψ_i is zero. This result, which is classical in the theory of vibrations and expresses a well known result in the mathematical theory of integral equations, here seems to come into fatal conflict with the conception of a u wave having a mathematical singularity. In fact, in the states where the singularity-particle ought to be immobile, it should be found at a point Q where the Ψ wave would be zero, that is, exactly at a point where, according to the statistical meaning of $|\Psi_i|^2$, it should be impossible to find it. The theorem expressed by formula (12) thus seems to force us to abandon my original idea of the u wave presenting a point-singularity like $1/r$, to wit, a function obtained from (5).[2]

3. The introduction of a non-linear wave equation for u

In order to avoid the difficulty that has just been pointed out, and for other reasons as well, we have been led to replace the idea of the u wave presenting a point singularity with another idea strongly sug-

[2] See conclusion of Section 6 for more detailed discussion.

gested by Vigier's remarks on the subject of the similarity between my demonstration of the guidance formula and the results of Georges Darmois and Einstein on the subject of the motion of a particle in General Relativity.

In General Relativity the coefficients $g_{\mu\nu}$, the space-time metrics, obey non-linear equations ($R_{\mu\nu} - \frac{1}{2}Rg_{\mu\nu} = 0$ in a vacuum), and the same is naturally true of the quantities $\gamma_{\mu\nu} = g_{\mu\nu} - g^{(0)}_{\mu\nu}$, which are differences between the $g_{\mu\nu}$ and their constant Galilean values $g^{(0)}_{\mu\nu}$. Nevertheless, except for the small singular regions of space-time which, according to Einstein, would make up the world-tubes of the particles and where the $\gamma_{\mu\nu}$ could take on very large values, the $\gamma_{\mu\nu}$ *approximately* obey linear equations; this property is very useful for the calculations made in the General Theory of Relativity. The chief result of Darmois and Einstein's researches is that these singular regions must move in the course of time in such a way that the extremely fine world-tube representing this motion coincides with a geodesic of the external field. This is a most remarkable result, because it permits us to deduce the motion of the particles directly from the field equations, without our being obliged to introduce as a special postulate (as is done in elementary presentations of General Relativity) the fact that the world line of a particle is a geodesic in space-time. The similarity of Darmois and Einstein's result to my demonstration of the guidance formula leads us to think that the "field" may very well be connected with space-time geometry and likewise obey a non-linear equation.

Let us see how this idea may be more clearly stated in the framework of the theory of the Double Solution. Obviously the wave equation for Ψ (which is a fictitious wave that simply expresses a probability) must be linear, for the principle of superposition, which is a necessary consequence of the statistical significance of Ψ, must be satisfied. The linear equation for Ψ is the one that is well known in usual Wave Mechanics. The theory of the Double Solution assumes that, except in a very small region constituting the "particle" in the strict sense of the word, the u wave obeys the same linear equation as the Ψ wave. But this does not keep us from assuming that the true u wave equation is a non-linear one, since the non-linear terms have no perceptible influence outside a very small, generally mobile, region in space where the values of u would become very large. Outside this small singular region,[3] the

[3] The adjective "singular" does not necessarily signify that u has, at a certain point in that region, a real point singularity. (See conclusion of the preceding chapter.)

non-linear terms would be negligible enough for u to obey *very approximately* the same linear wave equation as Ψ. If one adopts this new point of view, one is led, in order to demonstrate the guidance formula, to envisage a small sphere surrounding the singular region and forming its limits, namely, where the u function begins to increase rapidly but where it still obeys a linear equation.

The non-linear wave equation satisfied by u (the exact form of which could come out of Vigier's or similar attempts) is as yet unknown; it would certainly vary according to the nature of the particle and the value of its spin, and the values of the function (generally involving several u_k components) would determine what one may call the "internal structure" of the particle. This structure, in the case of particles of spin $\frac{1}{2}$ (in units of $\hbar$), such as the electron, would have a symmetry corresponding to this spin and be of the "dipolar" type. For particles of spin other than $\frac{1}{2}$, the symmetry would result, if the fundamental idea of my theory of "fusion" is correct, from a fusion of several elementary particles of spin $\frac{1}{2}$; there would then result a sort of confluence of the singular regions which would give rise to the appearance of symmetries other than the dipolar type (which is reduced to simple spherical polar symmetry in the case of zero spin). The dimensions of the singular region would doubtless permit us to define, at least approximately, a "radius" of the particle. We know that classical theories, like Lorentz', introduced such a radius, especially for the electron. Present theories very strongly feel the need of reintroducing the notion of the radius of a particle; but, having at their disposal for description of the particles only the statistical Ψ element, which does not allow an individual structure to be defined, they find themselves grappling with overwhelming difficulties.

I would now like to insist on a very important point. Einstein has emphasized that, if a field equation is linear, one can always find a singularity solution for it where the singularity has a motion prescribed in advance. For example, if, in the Lorentz theory of the electron, we wish to calculate the electromagnetic field created by a point-electron with a previously prescribed motion, one can always find the solution. In addition, always as a result of the linear character of the assumed equations, one can superimpose a continuous solution on the singularity solution without the former in any way affecting the latter. It is this that makes it necessary in Lorentz' theory, when one wishes to calculate the action of an electromagnetic field on an electron, to introduce as a supplementary postulate the existence of the Lorentz force exerted by the electromagnetic field on the electron. In order for the singular region to be automatically "guided" by the surrounding field, we must go outside the domain of linearity and base our theory on non-linear equations. It is because the field equations of General Relativity (that is, the equations for $g_{\mu\nu}$) are non-linear that Darmois and Einstein were able to find, without any supplementary postulate, a law for the guidance of particles by the field. By transposing these ideas to the causal interpretation of Wave Mechanics by means of the Double Solution, we see that, if we desire to establish between the particle and the wave a close bond that cannot exist in any linear theory, it seems quite natural to introduce non-linear equations of propagation.

One may even be led to conclude that, if present-day Wave Mechanics fails to give a clear explanation of the relationship of the wave and the particle, this failure results from the fact that it limits itself a priori *to the framework of a linear theory.*

However, in regard to what has just been said, one thing must be pointed out. The demonstration of the guidance formula that I gave in 1927 in no way seems to call for the non-linearity of the wave equations, even if still other reasons are found for assuming such non-linearity. This fact seems to contradict Einstein's considerations since, even though we assume that the u wave is a point singularity solution of a linear equation—as I did in my 1927 Paper—the guidance law imposes a well determined motion on the singularity. But one must note that, in order to make my demonstration, I assumed, in addition to the hypothesis of the linearity of the wave equation, the concordance of the phases of the u and Ψ waves, at least in the immediate region of the singularity. Further along we will see that, outside of the singular region, one must be able to write the u wave in the approximate form $u = u_0 + v$. In this formula, v designates a regular solution of the linear wave equation which, while it has an objective character, must *in general* be proportional to the Ψ wave considered in the usual interpretation. As for u_0, it is a peak-singularity solution of the linear wave equation which would possess a true point-singularity at the center of the singular region if the linear equation were still valid in that region. Now, if the wave equation were everywhere linear, u_0 and v would be totally independent solutions, and the postulate of the concordance of phases of u_0 and v would be a hypothesis arbitrarily tacked on. On the other hand, if the true u wave equation is non-linear, then, since the non-linear terms are essential in the singular region though negligible to all intents and purposes outside that region, the decomposition of u into u_0 and v can be approximately valid only outside the singular region. But, in reality, there is only one undecomposable u wave, since the two terms u_0 and v find themselves, in the singular region, entirely "fused" by non-linearity. So it becomes easily understandable why the postulate of the concordance of phases, which would be entirely arbitrary in a linear theory, may in the final analysis find its justification in the non-linearity of the equation of the u wave in the singular region.[4] Thus, in my 1927 demonstration, the hypothesis of a local non-linearity of the wave equation doubtless lay concealed in the postulate of the concordance of phases.

We will, moreover, soon see that there are still other reasons for adopting a non-linear equation for the propagation of the u wave.

4. The difficulty of determining exactly the relationship between the u and Ψ waves

Within the framework of the causal theory of the Double Solution, the u wave and the Ψ wave are of an entirely different character. The u wave must be an "objective reality"—that is to say, it must be independent of the observer and of the state of his knowledge.[5] The Ψ wave, on the other hand, is a probability representation of a subjective character dependent upon its user's knowledge and upon the information that observation and measurement have been able to supply to that user. This subjective character of the Ψ wave is especially obvious in the "reduction of the probability packet" which the physicist must effect when items of information (by supplying him with new knowledge concerning a particle or system) oblige him to modify his representation of the probabilities, and especially to eliminate from this representation whatever possibilities he now knows are not in fact realized.

[4] See, further on, formula (38).

[5] The fact that the u wave function is a complex quantity proves that it cannot directly represent a physical phenomenon such as the vibration of a medium; but this does not mean that it does not have an "objective" meaning, *i.e.* a meaning independent of the observer. That is the important thing from the point of view of the theory we are proposing.

Nevertheless, in spite of the essential difference in the nature of the u and Ψ waves, there must be a very close connection between their mathematical forms—a connection that is expressed by the equality of their phases and by the relation $\Box f/f = \Box a/a$ in the case of the Klein-Gordon equation, and also by the identity of the current-density vectors j_i arrived at by means of u and Ψ in the case of Dirac's equations. The difficulty, then, is to understand how such a connection is possible, and how it is compatible with the reduction of the packet of probability which, while quite comprehensible for the subjective Ψ wave, must not exist for an objective wave such as u. So we find ourselves confronted with the following very difficult problem: to establish a formal analogy between u and Ψ, *but in such a way that this analogy does not have as a consequence for the u wave a sharing of the subjective character of the Ψ wave.*

Let us examine in more detail a few special aspects of this general difficulty.

Let us first consider the case of a quantized system. The energies of its stationary states are defined as being the eigen values of Schrödinger's equation. Now these eigen values are obtained by proceeding from a linear partial differential equation and on the basic hypothesis that the Ψ solution (eigen function) are everywhere finite, uniform and continuous. But, if the Ψ wave is fictitious and if the objective reality is described by the u wave in which a singular region exists—if not a point-singularity—how can we justify the success of the calculation of the eigen values from the Ψ wave point of view? One ought to be able to make this calculation by utilizing only the u wave and without paying any attention to the statistical and fictitious Ψ wave. One can see that this must require some sort of analogy between the mathematical form of the u and Ψ waves, but one that nevertheless respects their intrinsic difference.

Let us now consider the propagation of a plane monochromatic wave with interference and diffraction phenomena. We know at the present time that these phenomena exist, not only for light, but for electrons and all other particles. How shall we interpret these phenomena if the continuous waves of the Ψ type are fictitious and if objective reality is described by a u wave having a singular region? In order to understand the problem more fully, let us base our reasoning on the classical case of the Young interference experiment, that is, of a flat screen pierced by two small apertures A and B. A plane monochromatic wave strikes the screen normally. If z is the variable calculated along the normal to the screen, the incident wave will be written, in the usual theory, in the form

$$\Psi = ae^{2\pi i\left(\nu t - \frac{z}{\lambda}\right)} = ae^{\frac{i}{\hbar}(Wt - pz)}. \tag{13}$$

Since the whole matter is Classical Optics, the wave, on leaving the screen, is calculated by considering apertures A and B as two small coherent sources of the same intensity and by then superimposing their effects. It is easily shown that in the vicinity of the axis of symmetry and far from the screen, the surfaces $\varphi =$ const. are ellipsoidal, and that the surfaces of equal amplitude are orthogonal hyperboloids. From this we deduce the position of the approximately rectilinear fringes which are actually observable on a screen placed normal to Oz in the region of interference. But, from the point of view of the theory of the Double Solution, we should, in order to really describe objective reality, replace the continuous Ψ wave by a u wave involving a singular region. In that case, in order for the singular region coming from the left to pass into the region to the right of the screen, it would have to go through one of the Young apertures, and then these two apertures no longer seem to play symmetrical roles, as they do in the classical calculations. It seems natural to suppose that the u wave diminishes rapidly in amplitude as one moves away from the singular region and that this amplitude sinks to very small values as soon as the distance from the singular region becomes macroscopic. If that were the case, the amplitude of u would be very great in one part of one of the apertures, while it would be very small over the entire surface of the other aperture; so it would thus seem impossible to admit that the two apertures play the symmetrical role of two sources of the same intensity. The hypotheses of the equivalence of the apertures, though seeming indispensable to the successful calculation of actually observed fringes, could no longer be maintained. The more one thinks about it, the more one has the impression that the difficulty is considerable.

Another series of difficulties is connected with the basic property of continuous wave-trains of the classical type, namely, their constant tendency to spread out in space with a constant diminution of their amplitude at every point (the latter property being tied in with the former because, the integral $\int a^2 d\tau$ being constant for continuous waves of the classical type, the extension of the wave-train has as its consequences a local diminution of a). On the other hand, the u wave, by the very fact that it must describe the objective structure of the particle, must possess a kind of permanence entirely different from the constant tendency to spread out, entirely different from the instability of the Ψ wave-trains. So we must expect to run into very great difficulty when we wish to establish a relation between the mathematical form of u and that of Ψ. We find here, in a new and rather different form, the objection that was formerly made to Schrödinger's conception, which sought to reduce the particle to a wave-group. That conception, which also ran into the objection that a discrete and well localized particle cannot be represented by a homogeneous and extended wave-train, had to be abandoned, because the constant expansion of a wave-train in the course of time—a basic property of continuous waves in linear propagation—is not compatible with the stability and prolonged permanence which the idea of a particle implies.

In a somewhat similar order of ideas, we also run into difficulties when we consider the fact that homogeneous wave-trains can be split into several trains of lesser amplitude—as happens when a wave-train is directed onto a semi-transparent mirror. For the continuous Ψ waves this effect is easily interpreted by considering each of the final wave-trains as representing the possibility of a final state whose probability is naturally less than the initial state (taken equal to unity). But it does not seem this could also be the case for the u wave which, describing the objective structure of the particle, does not appear divisible in this way. There are still other difficulties in the theory of the Double Solution which will arise when we come to the reduction of the probability packet affecting the Ψ wave, but which cannot affect the u wave. We will see that in this question the fact that the wave-trains are always limited plays a most important role.

In short, if we wish to determine the exact relationship between the u and the Ψ wave, we must expect to run into many obstacles. Yet that is what we shall attempt.

5. The external form of the u wave. The case of stationary states with an immobile particle

Vigier had suggested to me that the external part of the u wave outside the singular region might coincide with the Ψ wave. Presented in such simple terms, this idea struck me as one that would give rise to great difficulties, which do in fact exist and which I will discuss later;

and these difficulties caused me to reject Vigier's idea. However, I have reconsidered my position in a Paper of April 13th, 1953 [8]. That paper was the result of reflections on the first of the difficulties that I pointed out in the preceding section: How does it come about that, if the Ψ wave is purely fictitious and the u wave represents objective reality, the calculation of the eigen values of a quantized system succeeds when we start with the hypothesis that the wave is a finite, continuous and uniform solution of a *linear* wave equation? This seems paradoxical, since the u wave, which alone possesses an objective reality, does *not* have those very properties.

First, I will explain the considerations I was led to from reflection on this problem in the case of a stationary wave of a quantized system where the particle remains motionless (in the s state, for example). Then the Ψ wave has the form

$$\Psi_n(x, y, z, t) = a_n(x, y, z)e^{\frac{i}{\hbar}E_n t},$$

a_n being a real, finite, uniform and continuous function of zero value at the limits of the region defining the eigen-value problem under consideration. If one assumes that, in this stationary state, physical reality is described by a u wave of the form

$$u(x, y, z, t) = f(x, y, z)e^{\frac{i}{\hbar}E_n t}$$

the question is to find out why E is necessarily equal to one of the E_i, while f does not at all possess the properties of the eigen functions of a_n. To facilitate studying the question, I assumed—following a suggestion of Vigier's—that the u wave could be decomposed into a singular portion u_0, which would become very large but not necessarily infinite in the singular region and very small outside that region, and a regular portion v which would obey the linear equation of Wave Mechanics. This decomposition may appear arbitrary. I will take up shortly the reasons that may serve to justify and clarify it.

I will also assume, as seems quite natural, that u must be of zero value on the limits of the region. Then, except for the entirely exceptional case where the singular region of very small dimensions ($\leqq 10^{-13}$ cm) would be in the immediate vicinity of the limit of the region, one can consider v to be, to all intents and purposes, zero along the boundaries where $u_0 \simeq 0$. This also points up the fact that it amounts to practically the same thing to assume that u is zero on the boundaries of the region or to impose this condition on v. Let us represent the u

function as follows. (Naturally the problem is generally three-dimensional, which cannot be represented on paper, but this really makes no difference.)

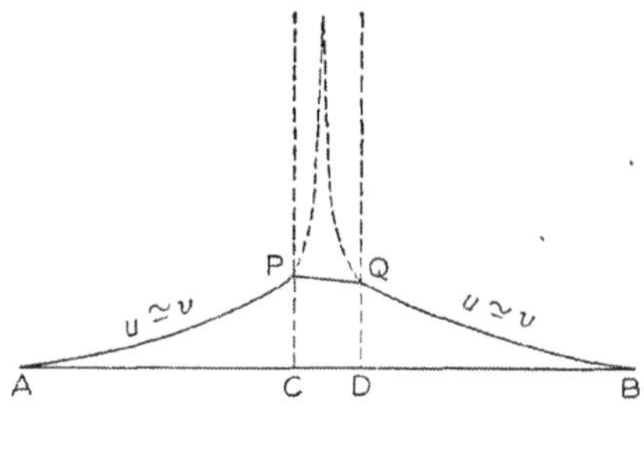

Fig. 10

Outside the singular region (immobile, by hypothesis) we have $u \simeq v$. Inside the singular region CD, the u function has very high values symbolically represented by the dotted curve. On the graph of u, let us consider two points P and Q where u still coincides approximately with v, and let us replace the u curve between P and Q with a portion of the PQ curve that is a finite, continuous and uniform solution of the linear equation. Then we have a curve APQB that will exhibit angular points at P and Q but that will be made up of finite sections of curves representing finite uniform and continuous solutions of the linear equation of Wave Mechanics. Such a "piecewise smooth" (*stückweise glatt*) function having zero values at limits A and B can be represented by a sum of eigen functions Ψ_n of the form $\sum_n d_n \Psi_n$. We see thus that outside the singular region, the u function may be written

$$u = u_0 + \sum_n d_n \Psi_n \simeq \sum_n d_n \Psi_n. \tag{14}$$

To be rigorous, one would have to point out that the Ψ_n are not exactly the eigen functions corresponding to the condition at the limits $\Psi_n = 0$, but rather the eigen functions corresponding to the limits $\Psi_n = -u_0$; but the two conditions at the limit differ by an infinitesimally small amount. We should then have *at every time*

$$f(x, y, z)e^{\frac{i}{\hbar}Et} = u_0 + \sum_n d_n a_n(x, y, z)\, e^{\frac{i}{\hbar}E_n t} \tag{15}$$

over the whole external region, and that is not possible unless

$$\mathrm{d}_n = 0 \qquad \text{for } n \neq m \qquad \mathrm{E} = \mathrm{E}_m \qquad u_0 = f_0(x, y, z)e^{\frac{i}{\hbar}\mathrm{E}_m t} \tag{16}$$

so that all the terms of (15) are periodic with the same frequency, E_m/h. Finally, one necessarily has

$$u = f(x, y, z)e^{\frac{i}{\hbar}\mathrm{E}_m t} \tag{17}$$

E_m being one of the eigen values calculated by ordinary Wave Mechanics. In this way we eliminate the paradox with respect to the calculation of eigen values and, at the same time, we see that outside the singular region we must have

$$u \simeq \mathrm{d}_m a_m e^{\frac{i}{\hbar}\mathrm{E}_m t} = \mathrm{d}_m \Psi_m. \tag{18}$$

The whole external portion of the u wave would thus coincide approximately, as far as the mathematical form is concerned, with the Ψ wave considered in ordinary Wave Mechanics; but here we must add a very important point: Since the u wave is an objective reality, it must have a perfectly determined value. Thus, if Ψ_m designates the mth *normalized* eigen function, d_m must have a well determined value. Outside the singular region, the u wave is approximately proportional to the Ψ wave considered by Wave Mechanics, but with a physically well determined coefficient.

Let us also point out that the foregoing reasoning proves that u_0 must, in the external region, obey a linear equation with $\mathrm{E} = \mathrm{E}_m$.

If we compare the ideas just developed with the demonstration of the guidance formula, we see that there is reason to distinguish three regions:

1) the "singular region" of radius r_0 ($r_0 \leqq 10^{-13}$ cm) in which the non-linear terms of the equation in u are significant;

2) an "intermediate region" defined by $r_0 < r < r_1$, where r_1 is likewise very small (doubtless of the same order of magnitude as r_0), in which the wave equation is already approximately linear, but where $u \simeq u_0 + \mathrm{C}\Psi$ increases rapidly as r diminishes by reason of the already rapid increase of u_0;

3) the "external region" which lies outside the sphere $r = r_1$, where the equation of the u wave is practically linear and where one may put $u \simeq v = \mathrm{C}\Psi$.

It is in the intermediate region that the sphere S, which serves in the demonstration of the guidance formula, must be placed; indeed, that demonstration assumes that on S the wave equation is linear, but that

u increases very rapidly as one penetrates into the sphere. Let us also note that the points C and D of Fig. 10 must correspond to $r = r_1$.

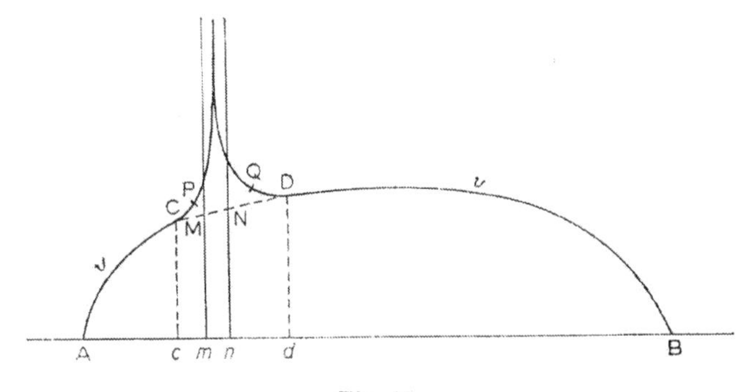

Fig. 11

Let us represent the ensemble of the three regions by another figure.[6] The very narrow region limited by the verticals Mm and Nn such that $\overline{mn} = 2r_0$ is the "singular region" where the non-linear terms are significant and where u takes on high values. The "external region" is, by definition, the one that is outside the verticals Cc and Dd of distance $\overline{cd} = 2r_1$; that is where $u \simeq v$. Finally, the "intermediate region" ($r_0 < r < r_1$) is included between Mm and Cc on one side and Nn and Dd on the other: the equation is still approximately linear, but the curve representing the variation of $u \simeq u_0 + v$ breaks away from the one representing v and begins to rise rapidly. It is in this intermediate region that the sphere S with radius corresponding, for example, to $\overline{PQ}$ must be placed. The region CcDd must be so small that the phase φ of the u and Ψ waves may there be considered as having the same value throughout the region. We have already called attention to the fact that, for particles of very high energy, this hypothesis may no longer prove valid—a circumstance that would impose a limit on the utilization of Ψ waves.[7]

6. An illustrative example of the external decomposition of the u wave

In order to show clearly what the decomposition of the u wave in the external region means—a decomposition into a singular part u_0 and a

[6] In Fig. 11, c and d ought to be symmetrical with respect to $\overline{mn}$.

[7] Note that for $r = r_1$, one has $f \simeq a$ and $\Box f/f \simeq \Box a/a$.

regular part v—we will consider a particularly simple and instructive example.[8]

Let us consider the problem of quantization for a spherical enclosure of radius R devoid of any field of force, and let us limit our consideration to the spherically symmetrical functions. The equation of the Ψ waves in the non-relativistic approximation is

$$\Delta\Psi = \frac{\partial^2 \Psi}{\partial r^2} + \frac{2}{r}\frac{\partial \Psi}{\partial r} = -\frac{2m}{\hbar^2}\mathrm{E}\Psi = -k^2\Psi \tag{19}$$

for which the general solution is

$$\Psi = \left[\mathrm{A}\frac{\sin kr}{r} + \mathrm{B}\frac{\cos kr}{r}\right] e^{\frac{i}{\hbar}\mathrm{E}t}. \tag{20}$$

Since the Ψ wave must remain finite at the origin, one must put $\mathrm{B} = 0$ and, in addition, since Ψ must be zero for $r = \mathrm{R}$, one must put $k = k_n = n\pi/\mathrm{R}$ with n an integer. The nth eigen function will then be

$$\Psi_n = \mathrm{A}\frac{\sin k_n r}{r} e^{\frac{i}{\hbar}\mathrm{E}_n t} \qquad \left(\mathrm{E}_n = \frac{\hbar^2 k_n^2}{2m} = \frac{n^2 h^2}{8m\mathrm{R}^2}\right). \tag{21}$$

Let us now introduce the idea of the Double Solution in the original form I gave to it, that is, assuming u to be endowed with a point-singularity like $1/r$ and everywhere, except where $r = 0$, satisfying the linear wave equation. In order to keep our calculations simple, let us suppose that the particle has an internal structure with spherical symmetry and that the particle is motionless at the center of the spherical enclosure. Then the u wave, which can only be a function of r, will obey the equation

$$\frac{\partial^2 u}{\partial r^2} + \frac{2}{r}\frac{\partial u}{\partial r} + k^2 u = \varepsilon\delta(r) \tag{22}$$

and the general form of Ψ, lacking the second term on the right, (given in (20)) leads us to consider the following solution for u—a solu-

[8] Let us note, moreover, that one obtains an even simpler example by considering the plane monochromatic wave that is set up along Oz. According to formula (5) of Chapter IX, one is then led to put

$$u(x, y, z, t) = \left[\frac{\mathrm{A}}{\sqrt{x^2 + y^2 + \frac{(z - vt)^2}{1 - \beta^2}}} + \mathrm{C}\right] e^{\frac{i}{\hbar}[\mathrm{W}t - pz]}$$

A and C having physically well determined values.

tion having a singularity $1/r$ at the origin:

$$u_0 = -\frac{\varepsilon}{4\pi}\frac{\cos kr}{r}e^{\frac{i}{\hbar}Et} \qquad \left(E = \frac{k^2\hbar^2}{2m}\right). \tag{23}$$

But we may add to u_0 in the expression of u any continuous solution we choose of the homogeneous form of equation (23). We will thus obtain for u a decomposition $u = u_0 + v$ similar to the one we considered in the last section. If we limit ourselves to the eigen functions with spherical symmetry, and if we define the eigen values k'_n (slightly different from the k_n) by the condition $u(R) = 0$, namely $\tan k'_n R = -\varepsilon/4\pi A$, we are led, in order to assure a single frequency for the whole phenomenon to adopt the solution

$$u_n = \left(-\frac{\varepsilon}{4\pi}\frac{\cos k'_n r}{r} + A\frac{\sin k'_n r}{r}\right)e^{\frac{i}{\hbar}E_n t}. \tag{24}$$

If we assume $\varepsilon \ll A$, which corresponds to the point-like character of the particle, we will have, as soon as we move the slightest distance from the origin,

$$u_n \simeq A\frac{\sin k'_n r}{r}e^{\frac{i}{\hbar}E_n t}.$$

It is then easy to distinguish three regions:

1) a very small spherical region ($r < r_0$) in the immediate vicinity of the origin where the singular term $(\cos k'_n r)/r$ is entirely preponderant;

2) an intermediate region ($r_0 < r < r_1$) where u_0, and consequently u, increase rapidly as r diminishes;

3) finally, the region outside a very small sphere of radius r_1 where u is effectively $Ae^{\frac{i}{\hbar}E_n t}(\sin k'_n r)/r \simeq C\Psi_n$. We can represent all this in Fig. 12.

But all these results, which appear to agree with those of preceding chapters, are satisfactory in appearance only; indeed, in formula (24) the constant A would be arbitrary; it would in no way be determined by the internal structure of the particle. We here come back to Einstein's observation, according to which, when we start out with a linear equation such as (22), we can always add any continuous solution whatsoever, of the homogeneous equation, to a solution involving a point singularity. So it is not possible to combine the terms u_0 and v, that is, to incorporate the particle into the wave. On the other hand, it does become possible if we replace (22) by a non-linear equation whose non-linear terms will be significant only in a very small singular region surrounding the origin and that will be approximately reduced to the usual linear equation outside of this region. Then we could still have as an approximate asymptotic form of u, outside the singular region, form (24), but ε and A will both be constants perfectly determined by the form of the u wave inside the singular region. One can clearly see here that it is the non-linear terms that assure the close association of the

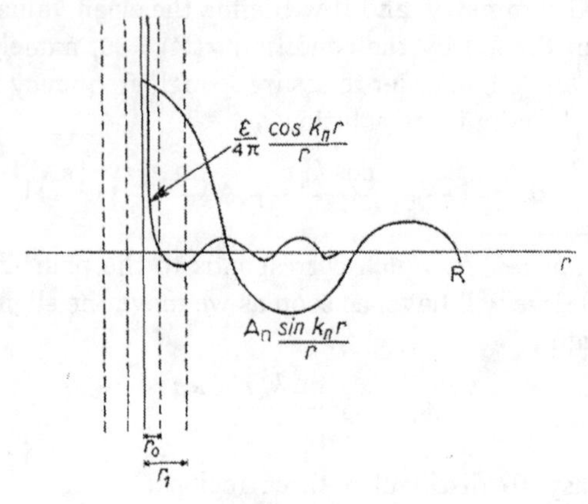

Fig. 12

singular wave u_0 and the regular wave v. One can say that it is in the non-linear region that the "fusion" of the sharply peaked u_0 wave and the surrounding regular v wave is effected.

Let us now return to formula (12) and note that, in the problem we are now considering, the function u_0, though very small on the boundaries of the enclosure, is not rigorously zero, and that the k'_n are slightly different from the $k_n = n\pi/\mathrm{R}$. Formula (12) must be valid for the function $u = u_0 + v$, a function with a point-like singularity becoming zero at the boundaries; but, since $k = k'_n$ does not coincide exactly with any of the k_n, it no longer contains any infinite term. In this way we overcome the difficulty that seemed to result from the Rayleigh-Sommerfeld theorem.

Let us examine the matter more closely. One can here write:

$$u(\mathrm{M}) = \sum_{i \neq n} \varepsilon \frac{\Psi_i^*(\mathrm{Q})\Psi_i(\mathrm{M})}{k_n'^2 - k_i^2} + \varepsilon \frac{\Psi_n^*(\mathrm{Q})\Psi_n(\mathrm{M})}{k_n'^2 - k_n^2}. \qquad (12')$$

Now

$$k_n'^2 - k_n^2 \simeq 2k_n(k_n' - k_n) \simeq 2k_n \frac{\varepsilon}{4\pi AR} = \frac{k_n \varepsilon}{2\pi AR}$$

and

$$\Psi_n(Q) = \left(N \frac{\sin k_n r}{r}\right)_{r=0}$$

since Q is at the origin. N is a normalization factor equal to $1/\pi \sqrt{k_n/2n}$.

One then finds, taking into account the value of the k_n,

$$\varepsilon \frac{\Psi_n^*(Q)\Psi_n(M)}{k_n'^2 - k_n^2} = A \frac{\sin k_n r}{r}$$

whence

$$u(M) = \sum_{i \neq n} \varepsilon \frac{\Psi_i^*(Q)\Psi_i(M)}{k_n'^2 - k_i^2} + A \frac{\sin k_n r}{r}. \tag{25}$$

Since $k_n = k_n' - \varepsilon/4\pi AR$,

$$\frac{1}{r} A \sin k_n r = \frac{1}{r} A \sin k_n' r - \frac{\varepsilon}{4\pi} \frac{\cos k_n' r}{R}$$

and we find

$$u(M) = \sum_{i=n} \varepsilon \frac{\Psi_i^*(Q)\Psi_i(M)}{k_n'^2 - k_i^2} - \frac{\varepsilon}{4\pi} \frac{\cos k_n' r}{R} + A \frac{\sin k_n' r}{r}. \tag{25'}$$

The first two terms give the expression for the function u_0, while the third one corresponds to the function v.

But the adding of v to u_0 in the expression of u with a well determined value of ε/A (which alone can give a well defined value of the k_n') could not be justified if the wave equation were everywhere linear.

Beside the very special case that we have just studied, similar considerations could be developed for all the problems of quantization that have a finite domain. If $\delta k_n = k_n' - k_n$ and $\delta\Psi = \Psi_n' - \Psi_n$ are the very slight variations undergone by the eigen values and the eigen functions when we pass from the classical limiting condition $\Psi = 0$ to the somewhat different limiting condition $u = 0$, we find for the general expression of $u(M)$

$$u(M) = \sum_{i \neq n} \varepsilon \frac{\Psi_i^*(Q)\Psi_i(M)}{k_n^2 - k_n^2} - \varepsilon \frac{\Psi_n^*(Q)\delta\Psi_n}{2k_n \delta k_n} + \varepsilon \frac{\Psi_n^*(Q)\Psi_n'(M)}{2k_n \delta k_n} \tag{26}$$

which contains (25′) as a special case. The first two terms of the preceding expression give u_0, the third gives v. But δk_n is not determined in a purely linear theory.[9]

7. Various extensions of the foregoing ideas

I have just explained the main idea of my Paper of April 13, 1953 for the case of a stationary state with a motionless particle. In that same Paper I generalized the result thus obtained in various ways.

A first very simple generalization concerns the case of stationary states where the particle is not motionless. Such is the case of the stationary states of the H atom whose Ψ wave contains the factor $e^{im\varphi}$ (φ being the azimuthal angle) and in which the particle possesses, according to the guidance formula, a uniform circular motion. The function Ψ_n is then written

$$\Psi_n = a_n(x, y, z)e^{-\frac{i}{\hbar}\chi_n(x,y,z)+\frac{i}{\hbar}E_n t}. \tag{27}$$

If we put

$$u = f(x, y, z, t)e^{-\frac{i}{\hbar}\chi(x,y,z)+\frac{i}{\hbar}Et} \tag{28}$$

and if, as we did previously and for the same reasons, we write the expression for u beyond the intermediate region

$$u \simeq u_0 + \sum_n d_n \Psi_n, \tag{29}$$

we will see that d_n must equal zero except for $n = m$, and E must equal E_n with, in addition,

$$\chi = \chi_m \quad \text{and} \quad u_0 = f_0(x, y, z, t)e^{\frac{i}{\hbar}(E_m t - \chi_m)}.$$

In other words, the phase φ of u and of u_0 must coincide with the phase of the Ψ_m and, outside the singular region, one will have approximately $u \simeq d_m \Psi_m$ where d_m will be a coefficient having a well determined physical value that could be calculated if one knew how to calculate u in the singular region.

Another immediate generalization consists of considering a quantized system that is in a non-stationary state, that is, a system whose Ψ function is a sum of eigen functions of the form

$$\Psi = \sum_n c_n a_n(x, y, z)e^{\frac{i}{\hbar}(E_n t - \chi_n(x,y,z))} = a(x, y, z, t)e^{\frac{i}{\hbar}\varphi(x,y,z,t)} \tag{30}$$

[9] For a specific example of the form that a non-linear equation for u might take, see ref. [17] pp. 56—59, where the u equation for a boson of spin 0 is given.

with the normalization condition $\sum_n |c_n|^2 = 1$. The considerations immediately preceding this section lead us to believe that the function v, a regular part of the u wave outside the intermediate region, is of the form $v = K\Psi$, where K is a constant with a well determined physical value. One can then postulate the expressions

$$u = fe^{\frac{i}{\hbar}\varphi'}; \qquad u_0 = f_0 e^{\frac{i}{\hbar}\varphi_0}; \qquad \Psi = ae^{\frac{i}{\hbar}\varphi}; \tag{31}$$

where the amplitudes and the phases are functions of x, y, z, t and, outside the singular region, we must have at every instant

$$fe^{\frac{i}{\hbar}\varphi'} \simeq f_0 e^{\frac{i}{\hbar}\varphi_0} + ae^{\frac{i}{\hbar}\varphi}. \tag{32}$$

We will be led to put $\varphi' = \varphi_0 = \varphi$. Outside the intermediate region, we will then approximately have $u \simeq Kae^{\frac{i}{\hbar}\varphi}$; thus the u wave will there be proportional to the normalized Ψ wave with a well defined coefficient of proportionality.

Having reached this point, we see appear the basic idea that the phases of the u wave and the Ψ wave must be the same—the postulate that is the very starting point of the theory of the Double Solution. This agreement in phase, which will be discussed again further on, here appears necessary in order for the function u_0 to be able to "mesh" with the v function that serves as its basis. The thing may be intuitively represented in the following manner: The higher values of the u function inside the singular region (which constitutes the particle in the strict sense of the word) forms a sort of needle or pinnacle that runs along the surface (or rather, in the very core) of the wave $v = K\Psi$ in such a way that it is always "in phase" with v. Naturally, it must not be forgotten that the needle in question is extremely fine, that its base has dimensions which are at the very most equal to 10^{-13} cm, and that (except for the case of particles of enormous energy) it can be considered almost point-like in proportion to the wave-length.

The result thus obtained is, then, most satisfactory, for it may be considered as the fulfillment of the basic idea that guided me in my earliest researches on Wave Mechanics—an idea according to which *the particle is a sort of clock that moves within a wave in such a way that it remains constantly in phase with it*—as I mentioned once more in the first chapter of the present work.

The same ideas may further be transposed to the case of a wave in free propagation. To be entirely rigorous, one must always consider such a wave as a limited wave-train. One then comes up against certain difficulties, which I will study in detail in the next chapter and which I pointed out in a Paper of August 17th, 1953 [12]. But we know that a long wave-train can be very approximately represented, along its whole extent except in the immediate vicinity of its boundaries, by a plane monochromatic wave. That is what makes it possible, in elementary treatises on Optics, to present all the arguments in terms of plane monochromatic waves although, in reality, one is always concerned with limited wave-trains. Proceeding, then, as one does in treatises on Optics, we will consider the particle as being associated with a plane monochromatic wave. The singular region runs through the core of this wave in such a way as to remain constantly in phase with it—a fact that obliges it to have exactly the velocity corresponding to the momentum $p = h/\lambda$ and to the energy $W = h\nu$. If the v wave which serves to maintain the singular region, happens to meet with obstacles that give rise to interference and diffraction phenomena in that wave, the singular region will have to shift, within the domain where these phenomena take place, in such a way as to remain constantly in phase with v; and this requires the guidance formula, as one may easily verify. One is then assured—with the reservation that the ergodicity theorem, which we have discussed at length, be demonstrated—that the probability of presence of the singular region at a point M in the field of interference at the time t is proportional to $|\Psi|^2$. In this way the existence of the particle is reconciled with the classical explanation of interferences.

In order to see this more clearly, let us consider once more the classical case of the Young apertures. The incident wave involves a singular region carried by a plane monochromatic wave v of the same form as the classical light wave of treatises on Optics. To say that the photon traverses the Young screen means that the singular region passes through one of the Young apertures, but the singular region is so small that, even at the very moment it goes through the aperture, it occupies only an infinitesimally small part of the aperture's surface, which has macroscopic dimensions. One can thus consider that, over the whole surface of the two apertures, the u wave constantly coincides with the v wave, that is, with the classical light wave. The interfering wave is thus exactly the same as the one considered in the classical explanation, with this single difference: we attach to it a local irregularity of extraordinarily small dimensions, an irregularity which is the singular region or particle and which runs through its core with the velocity prescribed by the guidance formula.

We now see why the interpretation of the Young apertures seemed to us, in Section 4, so difficult. It was because we thought of a u wave strictly localized around the particle (as the term u_0 is, in the expression obtained for u). If this were the case, the u wave would not be homogeneous over the whole surface of the aperture through which the photon passes, and it would have a negligible value over the surface of the other aperture, which is situated at a macroscopic distance from the first one. It would thus be impossible to understand how the classical calculation—which assumes a constant amplitude over both the apertures, whose roles are perfectly symmetrical—can produce a correct result. On the other hand, with our present conceptions, we see that everything takes place as it does in the classical interpretation, because the u wave extends far beyond the singular region in the form of a continuous wave that uniformly covers the two Young apertures and in which the singular region is, so to speak, implanted. In this way one of the most formidable obstacles to the adoption of the double-solution hypothesis—an obstacle that had at an earlier date contributed to my abandonment of it—seems to be removed.[10]

We have seen previously that, in the case of bosons, one can consider several bosons associated with one and the same wave. This means that the u wave of these particles is of the form

$$u = \sum u_0^{(i)} + v \tag{33}$$

outside the singular regions. The first term of the right-hand member of (33) represents the singular parts that are linked to the various bosons and which are extremely small outside their individual singular region. The term v represents the regular part of the u wave which is *common* to all the bosons, and that is what permits us to say that these particles are associated with one and the same wave.[11]

[10] One can consider the Young-aperture experiment as giving direct proof of the non-linearity of the u wave. In fact, v is propagated like a classical light wave, but if the equation of propagation of u were linear, the propagation of u_0 would be independent of v, and one could not explain how the particle's motion is influenced by the existence of a Young aperture which it does not go through. Only the non-linearity of the equation for u can have as a result the dependency of the propagation of u_0, *i.e.* the particle's motion, on the propagation of v. And it is because I did not introduce non-linearity in 1927 that I then thought that the Young-aperture experiment appeared to be an insurmountable obstacle to the theory of the Double Solution.

[11] For an amplification of the material in this section, see ref. [17] pp. 64—70.

8. The extension to Dirac's theory

The last extension indicated in my April 13th, 1953 Paper is one relative to Dirac's theory. All the results in the immediately preceding sections are obtained within the framework of a Wave Mechanics without spin and that is based on a non-relativistic wave equation or on the Klein-Gordon relativistic equation involving only a single Ψ. But the extension to the case of Dirac's theory for particles of spin $\frac{1}{2}$ may be made without difficulty.

The Ψ wave will then have four components Ψ_k and one will be obliged to write the expansions of Ψ as a series of eigen functions of the form

$$\Psi_k = \sum_n c_n \Psi_{n,k} \qquad (k = 1, 2, 3, 4) \tag{34}$$

$\Psi_{n,k}$ being the kth component of the nth normalized eigen function with the coefficients c_n not dependent upon the index k. Having put

$$v_k = \mathrm{K}\Psi_k = \mathrm{K}a_k e^{\frac{i}{\hbar}\varphi_k} \tag{35}$$

one will quite obviously be led to write

$$f_k e^{\frac{i}{\hbar}\varphi'_k} = f_{0,k} e^{\frac{i}{\hbar}\varphi_{0,k}} + \mathrm{K}a_k e^{\frac{i}{\hbar}\varphi_k} \qquad (k = 1, 2, 3, 4)$$

and to conclude from this, as previously, that $\varphi'_k = \varphi_{0,k} = \varphi_k$ for each of the four values of k. Outside the singular region one will have approximately $u_k = \mathrm{K}\Psi_k$, K being a physically well determined value resulting from the form of the u inside the singular region. Every u_k component is then, outside the singular region, proportional to the component of the Ψ wave having the same index. By reviewing what has been said in the previous chapter with regard to the guidance of the particle in Dirac's theory, one will see that the singular region must follow the lines of current of Dirac's theory and, from that point of view, the new conception of the relationship of the u and Ψ waves still appears satisfactory here.

9. Observations on the concordance of phases and the guidance of particles

Let us now once more take up the question of the concordance of phases, this time from our new point of view. As above, let us write

$$u = f e^{\frac{i}{\hbar}\varphi'} = u_0 + v = f_0 e^{\frac{i}{\hbar}\varphi_0} + \mathrm{K}a e^{\frac{i}{\hbar}\varphi}. \tag{37}$$

By examining the arguments presented above, we easily see that the conclusion $\varphi' = \varphi$ is required, but that the equality $\varphi_0 = \varphi$ does not in any absolutely necessary way result from them. One is not then obliged, it seems, to assume this latter equality, which constitutes the postulate of the concordance of phases in its strict form. It will be sufficient to assume that φ_0 and φ coincide, as well as their first-order derivatives, in the immediate vicinity of the particle, that is, in the "intermediate region" where it is necessary to situate the sphere S in demonstrating the guidance formula.

But in the external region, where u_0 is negligible in comparison with v, the phase φ_0 of u_0 is no longer important, and in the singular region, the decomposition $u = u_0 + v$ has no meaning. The distinction between a very local and a general concordance of phases is thus unimportant here.

The question may be put in another way. In reality, the decomposition $u = u_0 + v$ is fictitious. There is a single function u and that is all. Throughout the entire external region, that is, almost everywhere, u_0 is negligible and u coincides with v; u and v thus have almost everywhere the same amplitude and phase. In the intermediate region where the wave equation is, by hypothesis, still linear, u must always have the same phase as v, otherwise the guidance formula could not be correct; but the amplitudes of u and v differ, and this fact—calling the difference in amplitudes f_0—allows us to write for the periphery of the singular region

$$u = f e^{\frac{i}{\hbar}\varphi} = (f_0 + Ka) e^{\frac{i}{\hbar}\varphi}. \tag{38}$$

As for the form of u inside the singular region, it will remain unknown so long as we have not managed to find the exact form of the non-linear equation of propagation within that region. It is thus impossible for us to know at the present time whether, inside the singular region, the phase of u continues to be equal to φ; but that is not of great import here.

Naturally, similar considerations are applicable to Dirac's theory.

Let us now turn to another question that shows, as the preceding one does, the importance of the intermediate region where the coupling of u_0 and v takes place. Let us consider two particles that do not interact, two photons or two neutrons, for example. If their waves happen to intersect in the same region of space, it seems physically certain that they will not influence each other. If, in a light-interference apparatus, we make another beam of photons cross the field of interference transversely, the interference phenomenon will not be modified. This means that the wave of a particle can interfere only with other segments of itself and not with a wave of a different particle.

We can understand this point by analyzing what happens in the intermediate region bound to one of the particles, which we will label 1, while calling the other one 2. Except for the case—which is one of vanishing probability in all the usual cases of the crossing of beams in which two particles meet—*i.e.* in the case where their two singular regions would come in contact, we can write the u wave in the intermediate region of particle 1 in the form

$$u = u_1 + u_2 = f e^{\frac{i}{\hbar}\varphi} = f_0^{(1)} e^{\frac{i}{\hbar}\varphi^{(1)}} + \mathrm{K}a^{(1)} e^{\frac{i}{\hbar}\varphi^{(1)}} + \mathrm{K}a^{(2)} e^{\frac{i}{\hbar}\varphi^{(2)}}. \tag{39}$$

In fact, u_2 is reduced to v_2, since the intermediate region of 1 is in 2's external region. Since, in the intermediate region of 1, $f_0^{(1)}$ increases rapidly and becomes much larger than $a^{(2)}$, it is obvious that one has

$$\varphi \simeq \varphi^{(1)} \qquad f \simeq f_0^{(1)}. \tag{40}$$

The guidance formula then shows, by virtue of the first equation (40), that the presence of u_2 in no way modifies the motion of 1.

One can also see this by considering the formula

$$v = -\frac{\dfrac{\partial f}{\partial t}}{\dfrac{\partial f}{\partial s}} \tag{41}$$

which in Chapter IX, Section 6, permitted us to deduce the guidance formula. By introducing the second relation of (40) into (41), we see that the u_2 wave does not act on the motion of particle 1. In this form, the proof has the advantage of being applicable to Dirac's theory, as may be seen by going back to formula (53) of the preceding chapter.

These considerations seem to explain quite clearly why a wave interferes only with itself, and they also emphasize the importance of the way the "pinnacle" represented by u_0 imbeds itself in the intermediate region of the regular v wave.

10. The advantages of the foregoing conceptions. Difficulties still remaining

The foregoing conceptions make it possible to interpret the fact, paradoxical in appearance, that in the stationary states of quantized systems, the frequency of the u wave must be equal to one of the eigen values calculated by usual Wave Mechanics, even though the u wave does not possess the properties of regularity which this calculation imposes upon the Ψ wave. These same conceptions allow us to understand the possibility of interference and diffraction phenomena by reconciling the existence of localized particles and the classical calculations that permit us to predict them. In this way we overcome some of the most formidable difficulties that have been the stumbling blocks in all attempts to arrive at a clear representation of the wave-particle dualism.

Nevertheless, there still remain serious objections to overcome, notably with respect to wave-trains, their constant spreading tendency, and their breaking up in those cases where the probability-packet comes into play. Those were the difficulties that originally made me reject Vigier's hypothesis to the effect that the outer part of the u wave is proportional to the Ψ wave. The fate of the present attempt to arrive at a causal interpretation of Wave Mechanics most likely rests upon the possibility of resolving these difficulties in a satisfactory manner.

Let us conclude with an important observation that will play a certain role in the pages that follow. Since we assume the real equation of propagation satisfied by the u wave to be non-linear, the non-linear terms of that equation, although insignificant outside the singular region, nevertheless exist everywhere, in principle. If, then, these non-linear terms contained first-order derivatives, or derivatives of a higher order of u, these terms, although they will be negligible throughout the central portion of a wave-train, could none the less again become very important on the edges of the wave-train, where, because the u wave cancels out rather suddenly, its derivatives would have large values.

Chapter XVIII

WAVE-TRAINS AND THE REDUCTION OF THE PROBABILITY PACKET

1. A difficulty arising from the spontaneous spreading of wave-trains

A plane monochromatic wave is an abstraction; one is always dealing experimentally with wave-trains limited in space and taking a certain limited length of time to pass a given point. But we often deal with a wave-train which, throughout its whole central portion, excepting the regions near both ends, may be actually considered a plane monochromatic wave, as is the case for wave-trains utilized in Optics. Such a wave-train may be represented by a "wave-group", that is, by a superposition of plane monochromatic waves of very similar wave-lengths and directions of propagation. Now a wave-group so constituted has a natural tendency to spread out in space.

In order to see this, let us neglect the representation of the transverse dimensions of the wave-group by a superposition of plane waves having different directions of propagation, each of which is very slightly inclined with respect to the others; and, in order to represent the finite length of the wave-group in the direction of propagation x, let us write simply

$$\Psi = \int_{\mu_0-\Delta\mu}^{\mu_0+\Delta\mu} \mathrm{d}\mu\, c(\mu) e^{2\pi i(\nu t-\mu x)} \qquad \text{with } \mu = \frac{1}{\lambda}. \tag{1}$$

The frequency ν is a certain function of μ defined by an equation of propagation, assumed to be linear, which Ψ obeys, and this relation between ν and μ corresponds to the dynamic relation between energy and momentum. Let us put $\mu = \mu_0 + \eta$, $\nu_0 = \nu(\mu_0)$ and let us write the Taylor's expansion of $\nu(\mu)$:

$$\nu(\mu) = \nu_0 + \left(\frac{\partial \nu}{\partial \mu}\right)_0 \eta + \frac{1}{2}\left(\frac{\partial^2 \nu}{\partial \mu^2}\right)_0 \eta^2 + \ldots = \nu_0 + v_0\eta + \ldots, \tag{2}$$

where the index 0 indicates that the derivatives are taken at $u = \mu_0$ and where v_0 is the velocity corresponding to μ_0. We have made use of the fact that v_0 is given by Rayleigh's formula for the group-velocity, namely $v_0 = (\partial\nu/\partial\mu)_0$. So we may write

$$\Psi = e^{2\pi i(\nu_0 t - \mu_0 x)} \int_{-\Delta\mu}^{\Delta\mu} d\eta\, c(\eta) e^{2\pi i\eta(v_0 t - x)}\, e^{\left[2\pi i \left(\frac{\partial^2\nu}{\partial\mu^2}\right)_0 \eta^2 + \cdots\right]t} \tag{3}$$

where the dots represent terms of an order higher than η^2. The interval $\Delta\mu$ being very small by hypothesis and over a very long period of time beginning with the initial moment, the second exponential under the integral sign, whose exponent is approximately zero, may be taken as equal to 1, and we will have

$$\Psi = e^{2\pi i(\nu_0 t - \mu_0 x)}\, F(v_0 t - x). \tag{4}$$

The wave-group will then have the same phase-factor as the plane wave with frequency ν_0, and its amplitude will move as a unit with the velocity v_0 along the x-axis, and there is no spreading out.

But at the end of a sufficiently long time with respect to $1/\nu_0$, however small the interval $\Delta\mu$ may be, there will always come a moment when the exponent of the last exponential will cease to be negligible, and then the integral of (3) will be of the form $\int d\eta\, f(\eta, v_0 t - x_0, t)$ and we will have (except, however, in the case of a particle with zero rest-mass where $\partial\nu/\partial\mu$ is a constant)

$$\Psi = e^{2\pi i(\nu_0 t - \mu_0 x)}\, F(v_0 t - x, t). \tag{5}$$

The amplitude will thus become variable with the time in a way other than by dependence on the combination $v_0 t - x$. The wave-group will thus change its form as it goes forward, and a more detailed analysis would show that it will continue to spread out. Since the linear equations of propagation require the constancy of $\int d\tau\, a^2$ in the course of time, the spreading out of the wave-train has as a result a diminution of the local amplitudes, and thus there is a diminution of the wave-group that is spreading out.

This spreading-out of the wave-group may be intuitively explained in the following manner. The monochromatic waves, whose superposition forms the wave-group, are propagated *independently* of each other, because, by hypothesis, the equation of propagation of the Ψ wave is linear. Every monochromatic component has its velocity v corresponding to the value of μ that it specifies. Except in the case of particles of zero rest-mass where ν is a linear function of μ, some of the v velocities are greater and some smaller than v_0, the differences being in every case very slight. In the long run this means that certain components will "gain" on the central frequency component ν_0 while others will lag. Whence a spreading out of the wave-group, which, so to speak, breaks up slowly as it goes forward. This breaking up, accompanied by spreading out in space, is intimately tied up with the linear character of the equation of propagation.

We can anticipate this spreading out of the wave-group in still another way, which will be of use to us further on. Let us write the generalized Jacobi equation (J) corresponding to the linear equation of propagation of relativistic Wave Mechanics

$$\text{(J)} \qquad \frac{1}{c^2}\left(\frac{\partial \varphi}{\partial t}\right)^2 - (\text{grad}\ \varphi)^2 - m_0^2 c^2 = \hbar^2 \frac{\square a}{a}. \tag{6}$$

For the wave-group without deformation

$$\Psi = \mathrm{F}(v_0 t - x) e^{2\pi i(\nu_0 t - \mu_0 x)} \quad \text{with} \quad \frac{\nu_0^2}{c^2} - \mu_0^2 = \frac{m_0^2 c^2}{\hbar^2} \tag{7}$$

to be a *rigorous* solution of (6), $\square \mathrm{F}/\mathrm{F}$ would have to be equal to zero. That condition can be fulfilled for a form of F representing a wave-group of finite dimensions only if v_0 is equal to c, that is, for the case of particles of zero rest-mass. Thus, for a particle of non-zero rest-mass, the undeformed wave-group is not a solution of the linear wave equation. Let us note that for a wave-group having the form of Fig. 13, it is at the boundaries where F varies suddenly that the equation $\square \mathrm{F} = 0$ will not be satisfied.

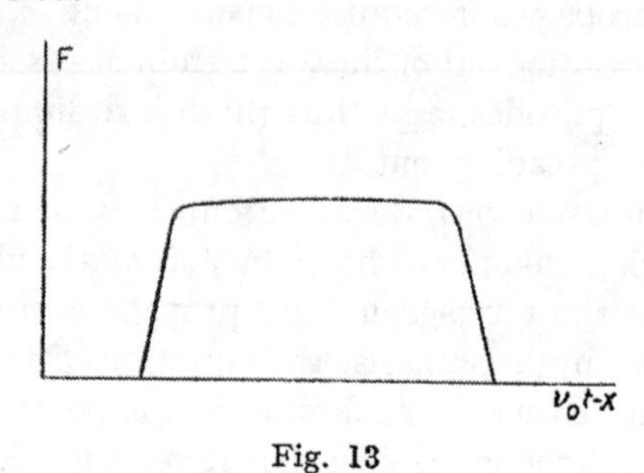

Fig. 13

Now here is the difficulty that arises when we compare these results with the conceptions introduced in the preceding chapter. If a particle is associated with a group of plane monochromatic waves, we ought, it seems, to admit that its u wave is quite closely represented, outside the singular region, by the expression

$$u = u_0 + v = u_0 + C \int \ldots \tag{8}$$

where C is a constant having an objectively determined value, and where $\int \ldots$ is the Fourier integral representing the wave-group. Unfortunately, at the end of a sufficiently long period of time, the wave-group will have spread out and dissipated. The regular v wave will then tend toward zero at every point and u will be reduced to its singular portion u_0. Figuratively, one may say that the particle will end up by "losing its wave", and that conclusion does not seem very acceptable, physically speaking. In short, in spite of the introduction of the "particle-as-a-singular-region-of-the-u-wave" hypothesis, which makes it possible for the theory of the Double Solution to preserve an objective meaning of the particle concept, we here run into the same objection that was formerly raised against the interpretation of Wave Mechanics proposed by Schrödinger, wherein the particles were said to be wave-groups.

2. The non-linearity of the wave equation might permit a conception of wave-groups without spreading

To overcome this apparently serious difficulty, one may ask whether it would not be possible to imagine groups of u waves that would not spread out. It is obvious that this could arise only from the difference between the u-wave and Ψ-wave equations, that is, from the non-linearity of the former.

Let us point out that, as we emphasized at the end of the preceding chapter, to put $u \simeq C\Psi$ outside the singular region is necessarily an approximation since, in view of the non-linearity of the u-wave equation, the non-linear terms, although very small outside the singular region, nevertheless exist *everywhere* and could again take on importance along the boundaries of the wave-train. In order to find out the exact "external" form of u, we would have to be able to evaluate the influence of the non-linear terms very exactly. It seems possible to conceive of the relation $u = C\Psi$ for $r > r_1$, as giving an exact representation of u in this domain when Ψ is a superposition of eigen functions belonging to a *discontinuous* spectrum (as was the case in the examples given in the preceding chapter) whereas it might give an erroneous representation, at least for certain regions, when Ψ is a Fourier *integral* representing a wave-group.

In other words, while in the case of a quantized system with a discontinuous spectrum, the outer part of the u wave could be quite closely represented by $u \simeq C\Psi$ (which would leave intact our interpretation of the success of the calculation of eigen values by the usual method), in the case of a continuous spectrum and of a group of Ψ waves, the outer portion of the u wave would, on the contrary, no longer be well represented *everywhere* by the corresponding Fourier integral. However, in the central portion of the wave-group, where we are not in the immediate vicinity of the wave extremities, the outer part of the u wave should indeed coincide very nearly with the "plane monochromatic wave" function, as that is necessary for the preservation of the interpretation of interference phenomena (of the Young-aperture type) that we obtained in the preceding chapter. So it would be *on the boundaries* of the wave-trains that the u wave might not be satisfactorily represented by the Fourier integral.

Let us pursue these ideas still further. First of all, the existence of the preponderant non-linear terms in the singular region must have the effect of "fusing" the two functions that we have called the "singular portion u_0" and the "regular portion v" of the u wave. This fusion ought to have the effect of tying the v wave inseparably to the singular region and thus keeping it from spreading away from it. We are once more back at Einstein's remark that the non-linear terms have the effect of rendering inseparable a regular solution and a singular solution of the field equations—solutions that would be independent if these equations were linear throughout. So it does seem that the intervention of non-linear terms in the singular region might produce a situation in which v is not exactly represented by a Fourier integral in the case of wave-groups, since such a representation implies the independence of the monochromatic components of the wave-group and, consequently, the spreading out of the wave-group in the course of time. But we must examine very closely what may go on at the boundaries of the wave-trains to prevent this spreading out.

We have already pointed out that the non-linear terms of the equation for u, negligible in the main part of the external region of the wave-train, may once again take on importance on the boundaries of that region. We have here a circumstance that may operate in such a way as to allow us to conceive of wave-groups without spreading.[1] To see

[1] It is interesting to note that a circumstance of this sort is generally ignored in the theory of the propagation of small gravitational disturbances in the General Theory of

that this is actually the case, let us write for the u wave the generalized Jacobi equation corresponding to the non-linear equation (in the absence of any field) as

$$\text{(J)} \quad \frac{1}{c^2}\left(\frac{\partial\varphi}{\partial t}\right)^2 - (\text{grad}\,\varphi)^2 - m_0^2c^2 = \hbar^2\frac{\Box f}{f} + \mathrm{N}\left(u, \frac{\partial u}{\partial t}, \frac{\partial u}{\partial x}, \frac{\partial^2 u}{\partial t^2}, \frac{\partial^2 u}{\partial x^2}\ldots\right)$$

where N is a non-linear expression, dependent upon u and its derivatives, which is to all intents negligible outside of the singular region except, perhaps, along the edges of the wave-train. Let us put, outside the singular region,

$$u(\mu_0) = \mathrm{F}(v_0 t - x)e^{2\pi i(\nu_0 t - \mu_0 x)} = \mathrm{F}(\vartheta)e^{2\pi i(\nu_0 t - \mu_0 x)} \tag{9}$$

with

$$\frac{\nu_0^2}{c^2} - \mu_0^2 = \frac{m_0^2 c^2}{\hbar^2} \quad \text{and} \quad \vartheta = v_0 t - x.$$

This form of u represents a group of monochromatic waves *without deformation.* It will be a solution of equation (J) if

$$\hbar^2\frac{\Box \mathrm{F}}{\mathrm{F}} = -\mathrm{N}\left(\nu_0\mu_0, \mathrm{F}, \frac{\partial \mathrm{F}}{\partial t}, \frac{\partial \mathrm{F}}{\partial x}, \frac{\partial^2 \mathrm{F}}{\partial t^2}, \frac{\partial^2 \mathrm{F}}{\partial x^2}, \ldots\right), \tag{10}$$

namely

Relativity (see von LAUE, *Die Relativitätstheorie*, vol. 2, p. 191 ff.). In that theory it is assumed that, in a small gravitational disturbance, the g_{ik} are of the form $g_{ik} = g_{ik}^{(0)} + \gamma_{ik}$ where the $g_{ik}^{(0)}$ are the constant Galilean values of the g_{ik} and where $\gamma_{ik} \ll g_{ik}^{(0)}$ are considered infinitesimals of the first order. One then writes for the g_{ik} the relation $\mathrm{R}_{ik} = 0$ valid outside of matter. In the expression of the R_{ik} we find the products of the classical quantities in General Relativity $\Gamma_{ik}^r = \left\{ {ik \atop r} \right\}$ which themselves are expressed in terms of the g_{ik} and their derivatives $\partial g_{ik}/\partial x^r$. One ordinarily considers these derivatives as infinitesimals of the first order, and the products of the Γ_{ik}^r are neglected as being infinitesimals of the second order. The non-linear equations $\mathrm{R}_{ik} = 0$ are then very approximately reduced (with a suitable choice of coördinates) to the linear equations $\Box\gamma_{ik} = 0$, and from this it is concluded that the very small gravitational disturbances are propagated in a vacuum with the velocity c.

But, as Chazy has emphasized (in his *Théorie de la Relativité*, Gauthier-Villars, Paris, 1928, vol. II, p. 148), it is not sufficient to assume in this demonstration that the γ_{ik} are infinitesimally small; one must assume that the derivatives $\partial\gamma_{ik}/\partial x^r$ are also infinitesimal. But these derivatives may become very large in the narrow regions of space-time corresponding to the boundaries of wave-trains if, in these regions, the γ_{ik} suddenly fall to zero. In such regions the linear equation $\Box\,\gamma_{ik} = 0$ is no longer valid and must be replaced by a non-linear equation. Thus, even in very small gravitational disturbances, there may be non-linear phenomena along the boundaries of the wave-trains.

$$\frac{\hbar^2}{F}\left(\frac{v_0^2}{c^2} - 1\right)\frac{d^2F}{d\vartheta^2} = -N\left(\nu_0, \mu_0, F, \frac{dF}{d\vartheta}, \frac{d^2F}{d\vartheta^2}, \ldots\right). \quad (11)$$

If N were everywhere negligible, it would be necessary to have either $v_0 = c$—which is possible only for particles of zero rest-mass—that is $d^2F/d\vartheta^2 = 0$, which cannot supply an acceptable form (limited at the two extremities) for the wave-group. But, by taking the non-linear terms into account, one sees the possibility of solutions of equation (J) which would represent limited wave-trains without deformation. In particular, a group of waves having the form represented in Fig. 13 could exist by reason of the intervention of the N terms on the edges of the wave-trains where the derivatives of F would be large.

It must be incidentally pointed out that F must also satisfy the continuity equation (C) which is here of the type

$$\text{(C)} \qquad \frac{\partial}{\partial t}F^2 + v_0\frac{\partial}{\partial x}F^2 + N'\left(\nu_0, \mu_0, \frac{dF}{d\vartheta}, \frac{d^2F}{d\vartheta^2}, \ldots\right) = 0$$

where N′ represents unknown non-linear terms arising from the equation of propagation of u. The F function being solely a function of $\vartheta = v_0t - x$, the sum of the first two terms of (C) will be zero and it will remain

$$N'\left(\nu_0, \mu_0, F, \frac{dF}{d\vartheta}, \frac{d^2F}{d\vartheta^2}, \ldots\right) = 0 \quad (12)$$

which imposes upon F a condition that will have to be compatible with (11), which means that there must exist a function $F(\vartheta)$ satisfying both (10) and (11).

It is easy to picture intuitively the possibility of these solutions representing wave-groups without deformation.

Let us write the equation for the u waves in the non-linear form and in the absence of any field, but limiting ourselves to the non-relativistic approximation:

$$\frac{\hbar}{i}\frac{\partial u}{\partial t} = -\frac{\hbar^2}{2m}\Delta u + N(u, \ldots). \quad (13)$$

Where the non-linear terms N are important, everything takes place as if there existed, in spite of the absence of any field, a sort of "potential barrier" represented by the N terms, a barrier that prevents the expansion of the wave-train. But this potential barrier is not imposed by any external action; it is created by the sudden variation of the u wave itself along the edges of the wave-train. By imagining simple forms for the N terms (the exact form of which is unknown), one can see that such "non-spreading" solutions must in fact exist.[2]

Nevertheless, one may wonder if the introduction of wave-groups without spreading, of the type envisaged above, is compatible with the Heisenberg uncertainty relations which are intimately bound up with the representation of wave-trains by a Fourier integral. On that score there does not seem to be as great a difficulty as one might at first think. In fact, the unspreading wave-group is represented by expression (9) with

$$v_0 = \left(\frac{\partial \nu}{\partial \mu}\right)_0 = \left(\frac{\partial W}{\partial p}\right)_0 \qquad \text{and} \qquad W = c\sqrt{p^2 + m_0^2 c^2}$$

($W \simeq (1/2m)p^2$ in the Newtonian approximation). Now, at a given instant, one can always expand the amplitude F in a Fourier integral of the form

$$F = \int d\eta\, c(\eta) e^{-2\pi i \eta x} \tag{14}$$

so that one can write, by putting $\mu = \mu_0 + \eta$,

$$u(\mu_0) = \int d\mu\, c(\eta) e^{2\pi i(\nu t - \mu x)} \qquad \text{with} \qquad \nu = \nu_0 + v_0(\mu - \mu_0). \tag{15}$$

At a given instant, one can thus conclude from this, by applying the usual reasoning, that, if the train of u waves has a length Δx, the interval Δp of the values of $p = h\mu$ in the integral will satisfy the inequality

$$\Delta x \Delta p \geqq h. \tag{16}$$

We thus obtain once more the Heisenberg relations with regard to the space variables.

Will we also again obtain the fourth Heisenberg uncertainty relation with regard to the time variable? What might at first make us doubt this is the fact that time enters differently in expression (15) of $u(\mu_0)$ and in the usual expression for the group of linear waves, the mathematical form of which is the same, but where

$$\nu = \nu_0 + v_0(\mu - \mu_0) + \tfrac{1}{2}\left(\frac{\partial^2 \nu}{\partial \mu^2}\right)_0 (\mu - \mu_0)^2 + \dots \tag{17}$$

So, in short, it is through the elimination—thanks to the non-

[2] The theory of these non-deformed wave-groups may be compared to the theory of "solitary waves" in Hydrodynamics, which exhibits certain similarities to it.

linearity of the equation in u—of all the terms of expansion (17) beginning with the third one—that is, by limiting ourselves to putting $\nu = \nu_0 + v_0(\mu - \mu_0)$—that we have avoided the spreading out of the wave-train. Now, by taking into account the relation between W and p, we see that formula (17) is written

$$\begin{aligned} \Delta W = W - W_0 &= v_0 \Delta p + \tfrac{1}{2} \frac{c^2}{W_0} \left(1 - \frac{p_0^2 c^2}{W_0^2}\right) (\Delta p)^2 \\ &= v_0 \Delta p \left[1 + \tfrac{1}{2}\left(1 - \frac{v_0^2}{c^2}\right) \frac{\Delta p}{p_0} + \dots\right]. \end{aligned} \tag{18}$$

Since, in a wave-group, Δp is always much smaller than p_0, it is clear that the bracket in the right-hand member is approximately equal to one. As for the unwritten terms, they are negligible in comparison with $v_0 \Delta p$ (and even exactly zero in the Newtonian approximation). Thus, in spite of the modifications that we have introduced into the expression of ν as a function of $(\mu - \mu_0)$ by passing from the wave-group with spreading to the wave-group without spreading, we still have the right to put

$$\Delta W \simeq v_0 \Delta p \tag{19}$$

and, since the duration Δt of the passage of a wave-group past a point in space is obviously given by

$$\Delta t = \frac{\Delta x}{v_0}, \tag{20}$$

we still have, by taking (16) into account,

$$\Delta W \Delta t \simeq \Delta p \Delta x \geqq h. \tag{21}$$

And that is the fourth Heisenberg uncertainty relation with its usual interpretation.

The analysis here given is, moreover, very instructive in regard to the passage from the usual wave-group to the wave-group without deformation. One might object, in regard to this passage, that by suppressing all the terms in the right-hand member of formula (17) beginning with the second, we modify the relation between energy and momentum in a way that deprives it of its character of relativistic covariance. It seems to us that the following reply may be made to this objection: The relativistic covariance of the relation between energy and momentum is defined in the framework of *Special* Relativity; now, by introducing non-linear terms of the General Relativistic type, we have in reality moved beyond the framework of Special Relativity.

Let us now pass from the case of a wave-group to that of wave-trains where Δp is not very small compared to p_0. In the usual representation, by means of Fourier integrals, of the wave-trains of Wave Mechanics, one is led to introduce the notion of proper differentials and to replace the Fourier integral by a sum of such differentials.[3] As Sommerfeld in particular has pointed out, the physical significance of the notion of proper differentials is that they represent a wave-group of finite dimensions while avoiding the introduction of the plane monochromatic wave, which is not only an abstraction but which likewise may not be normalized. By replacing the Fourier integral by a sum of proper differentials, one expresses the fact that the wave-trains are formed by the superposition, not of plane monochromatic waves, but of limited wave-groups. From the new standpoint assumed here by the introduction of non-linearity and wave-groups without spreading, it seems natural to define the wave-trains as a superposition of wave-groups without spreading of type (9), that is, to represent a wave-train by the expansion

$$u = \sum_{\mu_0} c(\mu_0) u(\mu_0) \tag{22}$$

the sum $\sum_{\mu_0}$ being extended to a sequence of values of μ_0, usually very close to each other. If the edges of the wave-groups are very sharp, one may consider the functions $u(\mu_0)$ as approximately orthogonal to each other. It thus seems, unless closer study should reveal otherwise, that we can apply to expansion (22) the reasoning usually followed in regard to sums of proper differentials and still obtain the Heisenberg uncertainty relations.

3. A weakening of the bond, thus far assumed, between the u and Ψ waves

We have previously stated the existence theorem which, at the outset of my researches on the Double Solution, seemed to me necessary for the justification of that solution. At that time I formulated the theorem thus: *To every Ψ wave considered by usual Wave Mechanics there must correspond a u wave with the same phase.* We are now in a position to criticize this statement and give it a more precise form.

[3] See, for example, Louis de Broglie, *Théorie générale des particules à spin*, 2 ed., Gauthier-Villars, Paris, 1953, Chap. I, sec. 4.

Let us note first of all that, since for us the Ψ wave is only an abstraction of a subjective and statistical nature, it is not logical to begin with the Ψ wave and make the existence of the u wave, which we consider to be an objective reality, dependent upon it. Clearly, it should be the other way around. In addition, for reasons that will presently become much clearer, there is reason *to weaken* the assumed bond between u and Ψ in order to keep the u from partaking of the subjective character of Ψ. Thus we will now say: *To every u wave that furnishes an objective description of a particle conceived as the center of a wave phenomenon, there may be made to correspond a Ψ wave of the usual type which, throughout the domain external to the singular region of the u wave, has* in general *approximately the same mathematical form as u, so that we may put $\Psi \simeq Cu$.*

In this new formulation we write the relation between Ψ and u in the form $\Psi = Cu$, instead of $u = C\Psi$, in order to emphasize clearly that the Ψ wave is a construction of the mind made by starting with the objective u wave. Moreover, we have introduced into the statement the words "in general" and underlined them, for we are now obliged to think that there may be limits to this correspondence. These limits are due essentially to the fact that the equation of propagation of Ψ is rigorously linear, while that of u is non-linear. We have just had a first inkling of this state of affairs on studying wave-groups, since we have been led to ask ourselves whether it is not necessary to replace, in the external expression of u, the Ψ wave-groups of the classical type with a function "without deformation" which could differ considerably from the Fourier integral, especially along the edges of the wave-group where non-linear phenomena are found.

We will again come upon this same idea of a possible limitation of the correspondence between the u and Ψ waves when we study successively the representation of the emission from a source by a divergent wave, and then the division of a wave-train by a semi-transparent mirror.

4. The representation of emission from a point-source by a divergent wave

As we previously pointed out, Francis Perrin has raised an objection to the interpretation of Wave Mechanics by the pilot-wave theory. When particles of like energy are emitted in an isotropic fashion around a point-source, Wave Mechanics assumes that this emission must be represented by the divergent spherical wave

$$\Psi = \frac{A}{r} e^{-ikr + \frac{i}{\hbar} Et}, \tag{23}$$

k being a function of E defined by the wave equation. The $|\Psi|^2$ diminishes proportional to r^{-2} as we move away from the source, and this expresses the isotropic dissemination of the particles around this source, and the weakening of the divergent wave corresponds exactly to the probabilistic interpretation of the quantity $|\Psi|^2$. But, if we assume, as the pilot-wave theory requires, that the particle is guided by the Ψ wave, it would then become necessary to assume that the particle is guided by a wave that becomes increasingly weaker as it moves away from the source. Since nothing keeps us from observing diffraction or interference phenomena at very great distances from the source, these phenomena would have to result from a reaction exerted on the particle by an infinitesimally weak wave. As Perrin pointed out, that is hardly conceivable.

Let us now adopt the point of view of the Double Solution. If we take literally the ideas set forth in the preceding chapter, they would lead us to say that the external portion of the u wave coincides very closely with the function Ce^{-ikr}/r, C having a well-determined value. But that also is unsatisfactory. Besides the difficulty presented by the physical meaning of the singularity at the origin, the u wave would spread out progressively in space; the particle would gradually lose its external u wave as it moved away from the source—a circumstance that can hardly be admitted, as we have said.

From the purely mathematical point of view one could answer, it is true that the reaction of the u wave on the particle is expressed by the quantum potential which, depending solely on equal values of $\Box a/a$ and $\Box f/f$ in the neighborhood of the particle, does not depend on the *absolute values of* a and f. In the case of Dirac's theory, an examination of the formulas previously obtained for expressing the guidance of the particle would make possible a similar reply. But from the physical point of view, such a way out appears quite unsatisfactory, for it makes a wave, which ceases to exist at the boundary, play a physical role. It seems to me hard to rest content with such a purely formal answer.

On looking at the matter more closely, one sees, moreover, that even with the usual interpretation, the wave $\Psi = e^{-ikr}/r$ cannot give an exact representation of each particle emitted. There is, first of all, the difficulty arising from the fact that the wave (23) has a singularity at the origin—a circumstance contrary to the conditions generally imposed on the Ψ waves and rendering, moreover, impossible the normalization of the divergent wave. Physically, it is certain that the Ψ wave associated with any one of the particles emitted by the source must have an outer and inner boundary, that is, must be represented not by a spherically divergent and indefinite monochromatic wave, but by a group of waves of that type. It is even possible that, instead of forming an entire spherical layer, the Ψ wave may also be laterally limited. So, even within the framework of the usually-assumed interpretation, one must consider the divergent spherical wave e^{-ikr}/r as being only a statistically average representation of the total isotropic emission from the source. Each individual emission ought to be represented by a group of spherical waves radially limited and perhaps having azimuthal limitations in width.

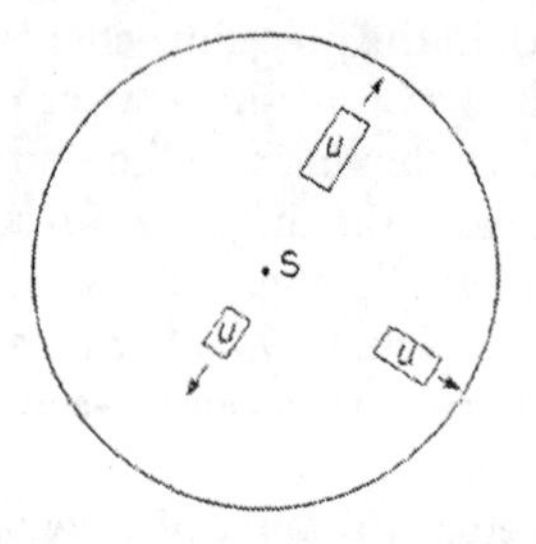

Fig. 14

If we then introduce the idea that there may exist, by reason of the non-linearity of the equation of propagation of the u wave, limited and non-spreading wave-groups, the source at S could emit isotropically in all directions wave-groups of a constant form (or approximately constant) each containing a singular region (particle).

Since the amplitude in each wave-group would then no longer diminish with r, the paradox pointed out by Francis Perrin would disappear. But we then see that the divergent wave $\Psi = (1/r)e^{-ikr}$ would, in this case, no longer be a wave associated with each particle, but simply a statistical representation of the overall, spherically isotropic, emission from the source.

Let us now make a few observations concerning the idea just put forward: the first one is that, in this hypothesis, the spherical and divergent Ψ wave would be purely fictitious and in no way bound in its totality with the groups of u waves emitted by the source. Its role would be merely to supply us with a representation of the statistical distribution, in the space surrounding the source, of the total number of particles emitted. From this point of view it would be somewhat like the Ψ wave of a system in configuration space—a wave which likewise, from our point of view, represents only the statistical behavior of particle localizations, and not at all the objective wave phenomenon that surrounds each singular region.

It will be noted further that this conception of the emission from a source of wave-trains with azimuthal limits, of which only the statistical distribution is represented by the divergent spherical wave, is quite similar to the idea of "needle" radiation (*Nadelstrahlung*) put forward not too long ago by Einstein in an effort to represent the luminous emission from a source, photon by photon, with each individual emission to be accompanied by a "recoil" of the source.

Finally, let us note that the hypothesis, according to which the divergent spherical wave representing the emission of photons or of material particles by a source is only a fiction, does not contradict the well-established existence of the phase-change of π that accompanies the passage of a spherical wave through a focal point. In fact, this phenomenon is observed when there is a passage through a focus of the wave-train associated with a *single* particle. In that case, the classical mathematical form of the Ψ wave that converges towards a focus and then diverges from it must represent the regular external region of the particle's u wave, and that allows us to account for the change in phase at the time of passage through the focus. The case of the divergent wave representing emission from a point-source is, moreover, entirely different from the one just considered. In fact, the divergent spherical wave has a singularity at the point-source; and this results in the non-zero value of the "particle-current" vector through a small spherical surface surrounding the source—a fact which mathematically symbolizes the hypothesis of an emission by the point source. On the other hand, when one constructs the theory for the passage of a convergent wave through a focus,[4] one is careful to choose, for representation of the phenomenon, the solution of the spherical-wave equation which will remain *finite* at the focus, that is, $(\sin kr)/r$ and not $(1/r)e^{-ikr}$; and this is

[4] See, for example, Henri Poincaré, *Théorie mathématique de la lumière*, t. II, p. 163.

done so that there is a zero flow of particles through a spherical surface surrounding the focus, since this focal point is neither a source nor a sink for the particles.

Summing up, we see that, within the framework of the theory we have been explaining, the study of the divergent spherical wave, as of wave-groups, has suggested the following idea: The correspondence that we initially postulated between the regular part of the u wave and the Ψ function of Wave Mechanics must not be considered as being of an absolutely strict and general nature. We are going to find the same idea again when we study the division of a group of waves by a semi-transparent mirror.

5. The division of a wave-group by a semi-transparent mirror

First let us recall a few considerations that Heisenberg put forward concerning semi-transparent mirrors when the discussion of the interpretation of Wave Mechanics first got under way.

Let us consider, from the point of view of the Ψ waves, the particle's incidence on a semi-transparent mirror M. The incident Ψ wave is split into a transmitted wave and a reflected wave, which represent respectively the two possibilities open to the particle: either of being transmitted or reflected.

Having adopted the purely probabilistic interpretation of Wave Mechanics, Heisenberg, at the Solvay Congress of October 1927, made the following remarks: According to him, one should not say that, on reaching the mirror, the particle "makes a choice" between the reflected beam and the transmitted beam, for the arrival of the particle at the mirror is not an observable fact; on the contrary, so long as the particle has not been localized by an observation, one should say that it exists "in a potential state" in both the transmitted and the reflected wave. If at any given moment we succeed in determining the particle's presence in one of these beams, the other immediately ceases to exist because it corresponds to a possibility that is never realized, and this fact clearly indicates the non-objective character of the Ψ wave. But, said Heisenberg, if instead of seeking to localize the particle in one of the beams, we had at M′ a perfectly reflecting mirror, we could obtain interference in the shaded areas of the figure where the two beams overlap. There would then be variations in the probability of localization of the particle in that region; and this fact illustrates that, until we have localized the particle, we must consider the reflected and the transmitted beam simultaneously. That is the present-day purely probabilistic position on this question.

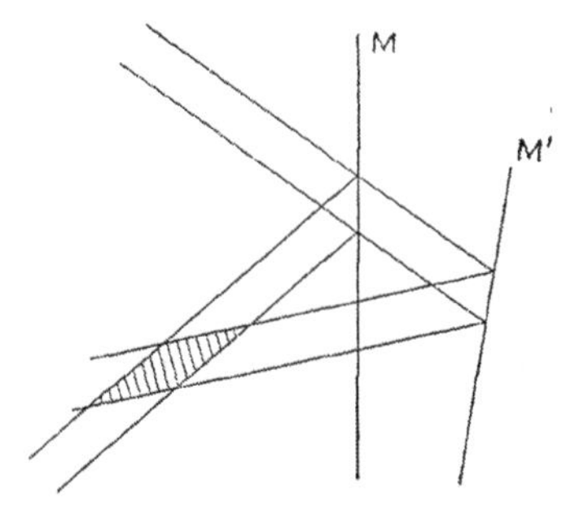

Fig. 15

Let us analyze a little more closely what happens when the semi-transparent mirror M splits up a wave-train. In the initial state we assume an almost monochromatic wave-group R_0 moving towards the

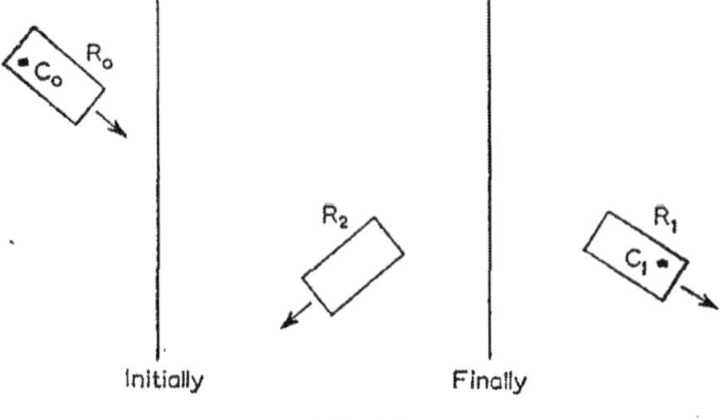

Fig. 16

mirror. In classical wave theory the presence of the mirror has the effect (after a very brief period during which the wave-group penetrates the mirror) of separating the incident wave-group into two wave-groups occupying regions R_1 and R_2 which are symmetrical with respect to the surface of the mirror and equal to R_0.

If the coefficient of reflection of mirror M is exactly $\frac{1}{2}$, and if in the initial wave one has $\int d\tau\, a^2 = 1$—the constancy of which integral in the course of time is assured by the equation of propagation—one will have in the final state within R_1 and R_2 an amplitude $a/\sqrt{2}$ in such a way that

$$\int_{R_1} d\tau \left(\frac{a}{\sqrt{2}}\right)^2 + \int_{R_2} d\tau \left(\frac{a}{\sqrt{2}}\right)^2 = 1.$$

If we then place the perfectly reflecting mirror M′ as in Fig. 15, the reflected and transmitted waves of amplitude $a/\sqrt{2}$ will interfere in the shaded area. On this point classical wave theory and the theory of Ψ waves agree, although they interpret the physical meaning of the interference in the shaded area differently. But if, instead of placing the mirror M′ as shown, we had made an observation that would have permitted us to localize the particle in one of the beams, in R_1 for example, we ought to say, after that localization, that there is no longer any Ψ wave in R_2 (for there is no longer any possibility of localizing the particle in R_2) and we ought to renormalize the Ψ wave in R_1, once again assigning to it the amplitude a instead of $a/\sqrt{2}$ in such a way as to have $\int_{R_1} d\tau\, a^2 = 1$. Naturally, this is meaningless in the classical theory of continuous waves for, in that theory, waves have a physical meaning, and an observation of R_1 can in no way modify the wave existing in R_2. In this way the inherent difference between the classical continuous wave of objective character and the Ψ wave of usual Wave Mechanics which, being merely a representation of probability, has a subjective character contingent upon the information we possess, is clearly emphasized.

6. The same problem considered in the theory of the Double Solution

Let us now introduce the theory of the Double Solution, involving, as it does, the u wave. We must assume that at every instant the particle has a well defined position in space, even if we have made no observation actually permitting us to localize the particle. Starting from an initial position in R_0, the particle will finally occupy a certain position in R_1 or R_2, the position C_1 in R_1 (Fig. 16) for example. So it is natural to think that, in its final state, the u wave whose singular region surrounds the point C_1 has a regular external portion that fills the region R_1. But is there a fraction of the external part of u which passes into R_2 and which, as a consequence, in the absence of the mirror M′, is then indefinitely separated from the particle, thus forming an isolated wave-group *devoid of singular region*? At first this idea struck me as quite unsatisfactory, and for a brief moment I considered the following hypothesis: In the classical picture, at the moment when the action of the mirror splits the incident wave-group into two *separate* wave-groups, the u wave would pass entirely into one of the wave groups (R_1 for example), and the other would lack any u wave and would simply represent—so long as we had no information on the particle's position—the possibility that the particle might have arrived in R_2. I will explain why it now seems to me that this hypothesis must be rejected.

Let us begin by recalling what the apparatus known as the Michelson interferometer is.

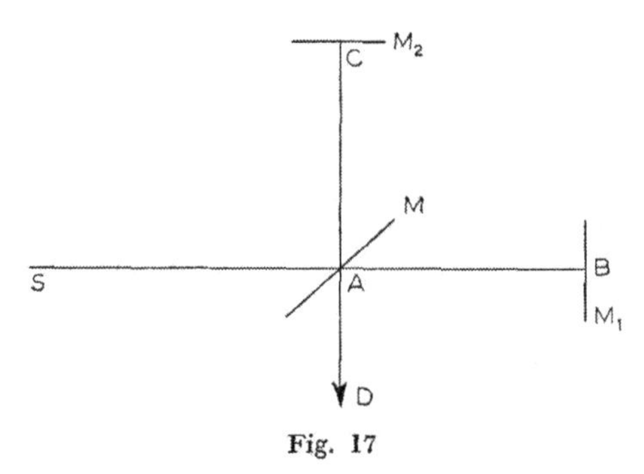

Fig. 17

A parallel light-beam strikes a semi-transparent mirror M at A, the mirror being tilted at an angle of 45° with respect to SA. The transmitted and reflected beams are then reflected, respectively, on the non-transparent mirrors M_1 and M_2. When the beam AB again reaches A, it is partially reflected by the mirror M in the direction AD, while beam AC, having once more reached A, is partially transmitted in the same direction AD. One can thus observe at D interference patterns that correspond to the possible difference in path of the rays SACAD and SABAD. The apparatus, as is well known, has the utmost precision.

But the situation which now arises is the following: If the incident wave-train is very long by comparison with the dimensions of the interferometer, it will, for a very brief period, occupy the entire apparatus (Fig. 18 (a), shaded area).

The wave-train will thus assume the form of a train of fragmental waves, but forming a single bloc. If, on the other hand, the wave-group is very small by comparison with the interferometer, after the group has been divided by the action of mirror M, it will give rise to two small wave-trains that are entirely separate and that cover paths ABA and ACA independently, meeting subsequently and interfering with each other in the direction AD (Fig. 18 (b),).

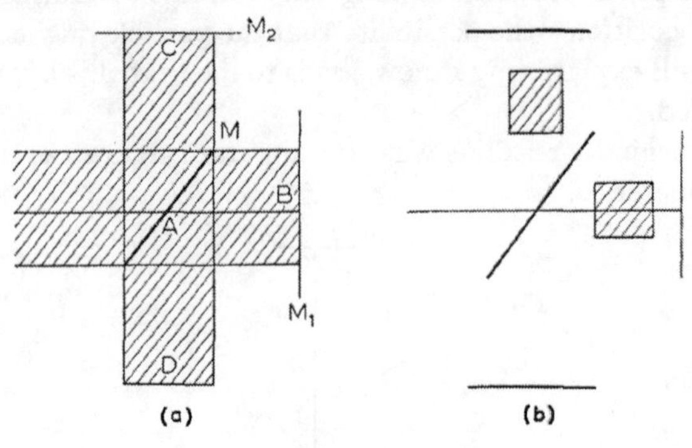

Fig. 18

Classical wave theory predicts that in both cases there will be interference at D. On the other hand, the *u*-wave theory, if we assume the hypothesis made above, would lead us to say that there might be interference patterns in the first case, because there would be no dislocation of the incident wave-train and because everything would take place within one and the same wave-group, whereas in the second case there would be no interference because there would then be a complete separation in space of the two wave-trains, one of which, devoid of the *u* wave, would not actually exist, since it would merely represent an unfulfilled possibility. Now such a conclusion, which is already *a priori* rather unlikely, is in flat contradiction with experience, as Renninger in particular has pointed out to me. In fact, Michelson and Gale have obtained interference patterns with wave-trains separated in space by distances of the order of two kilometers.[5] So we must reject the idea that, through the action of a semi-transparent mirror, a wave-train splits into two wave-trains separated in space, with the *u* wave entirely contained in the wave-train into which the particle has passed, thus leaving the other wave-train entirely devoid of *u* waves. We must reject it because the second wave-train would then be physically nonexistent and could produce no interference with the other wave-train when these two subsequently meet.

[5] "The Effect of the Earth's Rotation on the Velocity of Light," *Nature*, 113 (1923) 556.

These considerations seem, of necessity, to lead us to the following conception: When a wave-train is split in two by the action of a semi-transparent mirror, the u wave is shared by the two wave-trains in such a way that we finally have, on the one hand, a train of u waves in which a singular region, which is the particle, is embedded, and on the other, a train of u waves *without* a singular region. The new idea here appearing is that there exist u wave-trains without a singular region. Moreover, this idea appears to be in accord with the present formalism of "occupation numbers" in quantum field theory. According to that formalism one may have zero or one particle in the case of fermions; $0, 1, \ldots, n$ particles in the case of bosons. In both cases the occupation number may be zero, and that corresponds nicely to our conception of u wave-trains without a singular region.

It does not seem that the conception of u waves devoid of a singular region or particle presents, in itself, any difficulties.

Certain difficulties, however, do appear when one closely examines what happens in a case like that of the semi-transparent mirror in Fig. 16. In order to make our argument as precise as possible, let us assume that the reflective power of the mirror is equal to $\frac{1}{2}$. What we have just said leads us to assume that, if the particle finally reaches C_1 in the R_1 wave-train, there is nevertheless a portion of the u wave, everywhere regular, which has passed into the R_2 wave-train. Now the amplitude a of the Ψ wave and the amplitude f of the u wave (outside the singular region) satisfy the continuity equations

$$\frac{\partial}{\partial t} a^2 + \text{div}\,(a^2\,\mathbf{v}) = 0, \qquad \frac{\partial}{\partial t} f^2 + \text{div}\,(f^2\,\mathbf{v}) = 0, \tag{24}$$

$\mathbf{v}$ being, in both equations, the *same* definite velocity defined at every point by the guidance formula. By eliminating div $\mathbf{v}$ between these two equations and by introducing the total derivative $D/Dt = \partial/\partial t + \mathbf{v} \cdot \text{grad}$ taken along a line of current, one easily shows that $(D/Dt) \log f/a = \text{const.}$, that is to say that the quotient f/a must remain constant when we move along a line of current with the velocity $\mathbf{v}$. Now, starting out from any point in R_0, and following the line of current that passes through that point, we finally reach a certain point in R_1, or a certain point in R_2, depending on the initial position chosen in R_0. Since, according to our hypotheses, $a = Cf$ in R_0, it seems that one may deduce that at every point in R_1 or R_2 (except in the singular region surrounding C_1) the ratio f/a has the same value C that it initially had in R_0. Now we know that according to the usual linear theory of the Ψ wave, a must decrease by a factor $1/\sqrt{2}$ when we pass from R_0 to R_1 or R_2. Thus the amplitude f ought likewise to decrease by the same factor. But that conclusion precipitates another difficulty.

If, in fact, we are willing to admit that the passage through the semi-transparent mirror has the effect of weakening the external regular portion of the u wave, then successive passage through a very great number of mirrors will have the effect of virtually reducing the external portion of u to zero. Once more we arrive at the conclusion that the particle will progressively lose its outer wave—a conclusion that is difficult to admit from the physical point of view.

It would be far more natural, within the framework of our conceptions, to assume that when a particle has passed through a whole series of semi-transparent mirrors it has exactly the same properties it had when it was still near to the source. The problem is, in short, to know if a particle—whether it is associated with a wave-group in free propagation or whether is it emitted by a distant point-source or has passed through semi-transparent mirrors—retains invariable structure and properties, or whether, on the contrary, its structure and properties change—in other words, whether it "grows old". Although it may be possible to maintain the latter hypothesis, the former nevertheless appears to be much more likely. But the former one implies that the u wave of the particle, once it has arrived in R_1, should have the same external portion as it originally had in R_0 with the same amplitude f. But is that not irreconcilable with the result that we demonstrated previously with the help of equations (24)?

What might get us out of this dilemma is the fact that in reality the second equation of (24) is not rigorously valid if the equation of the u wave is not linear. The separation of the real and imaginary terms in the equation for u, when we put $u = fe^{\frac{i}{\hbar}\varphi}$, leads to a generalized Jacobi equation and a continuity equation, both containing a right-hand member made up of non-linear terms. It is this circumstance which, in the case of the generalized Jacobi equation, led us in Section 2 to consider the possibility of wave-groups without deformation. As a result of this same circumstance, the second continuity equation (24), which is rigorously of the form $(\partial/\partial t)f^2 + \mathbf{v} \cdot \text{grad } f^2 = N'$ could cease to be valid along the edges of the wave-train and, in principle, is never rigorously valid inside the singular region. This fact has as a consequence that one can no longer affirm that f/a has the same value in R_1 and R_2 as in R_0. One can also say that the non-linear terms of the equation for u make it possible for the region immediately surrounding the singular region, and also the boundaries of the wave-train, to behave as *sources* (or sinks) of the u wave. In the stable undeforming wave-trains, which we considered in Section 2, these sources should cancel each other out; but, when a wave-group of this type is split apart to give rise to two distinct wave-groups R_1 and R_2, these sources could reappear and cause the amplitudes of v to vary in R_1 and R_2. These amplitudes would be equal at the outset and remain equal for a certain length of time during which interference between R_1 and R_2, in the case of subsequent superposition, would still be possible under the conditions imposed by the classical theory. But after a certain lapse of time the action of the sources might produce a *regeneration* of the original wave-group in R_1. It might also be possible that the wave-group R_2, from which the particle is absent, might eventually disappear, perhaps through indefinite spreading. These are all merely suggestions, the undoubtedly difficult justification of which would be possible only if we succeeded in determining the form of the non-linear terms in the equation of the u waves.[6]

7. Reconsideration of the relationship between the u and Ψ waves

The study of the passage of a particle through a semi-transparent mirror, no matter what approach is finally developed in the theory of the Double Solution, will allow us to make clearer certain connecting links that should exist between the u and Ψ waves. In the original wave-group R_0, since the normalized Ψ wave has an amplitude a_0 (such that $\int d\tau\, a_0^2 = 1$), the relation between the Ψ wave and the external portion of the u wave is $\Psi = C_0 v_0$ with a constant C_0 of well determined

[6] Without giving undue importance to it, one may give an illustration of the idea here expounded by recourse to a metaphor suggested to us by J. L. Destouches. To a certain extent, one may compare a particle in uniform rectilinear motion surrounded by its wave-group to a ship moving on the open sea and surrounded by its wake. If the ship passes along a wall—a dike, say—a part of its wake may be reflected and, if it overlaps with another portion of the original wake, the reflected part may give rise to interference phenomena. But when the ship again moves away from the dike, it finds itself once more in the open sea with its original wake restored, whereas the portion of the wake reflected by the dike will have spread out into the open sea and disappeared.

value. After the separation of the two-wave-groups R_1 and R_2 equal to R_0, the particle is in one or the other of these two groups, without our being able to say which one, so we must imagine a Ψ wave filling both wave-groups with, as we have seen, the amplitude $a/\sqrt{2}$. If $|v_1|$ and $|v_2|$ represent the amplitudes of the external parts of u in R_1 and R_2, one will have for these two wave-trains respectively

$$a_1 = C_1|v_1| \quad \text{and} \quad a_2 = C_2|v_2|$$

with

$$C_1 = \frac{a_0}{|v_1|\sqrt{2}} \quad \text{and} \quad C_2 = \frac{a_0}{|v_2|\sqrt{2}};$$

thus, when we go from R_0 into R_1 or R_2, the constant of proportionality varies. If, then, we know for a fact that the particle is in R_1, we must renormalize the fictitious probability wave in R_1 to unity, and this will restore the wave's original amplitude a_0, whereas in R_2, we must put $\Psi = 0$, since we now know that the particle is not to be found in R_2. Naturally, the u wave which, for us, is an objective reality, cannot be modified either in R_1 or R_2 by the information that we obtain about the position of the singular region. So we must now put, in R_1, the relation $\Psi = C_1'v_1$, with $C_1' = a_0/|v_1|$, which is to say that the constant of proportionality C_1 must be multiplied by $\sqrt{2}$. But in R_2 one must put $\Psi = 0 \times v_2$, which means the constant of proportionality here becomes zero.

So it becomes quite clear how we must relax the possible bond between the regular portion of the u wave and the Ψ wave in order to make the objective character of the u wave compatible with the subjective character of the Ψ wave. In the train of waves in which the particle is finally situated, the u wave still remains proportional to the external part of the Ψ wave, but the constant of proportionality changes when there is a reduction of the probability packet and a renormalization of the Ψ wave. In the wave-train from which the particle is absent, one has to imagine, before any localization of the particle is determined, a Ψ wave that represents its probability of presence; but after the localization of the particle in R_1, there remains in R_2 only a u wave, regular throughout and *devoid of any Ψ wave.* These considerations seem to us to make much clearer the nature of the connection that may exist between the objectively real u wave independent of all our information, and the Ψ wave, which is a mental construct and a subjective representation of the probabilities which depends essentially upon the information we possess. Earlier, we pointed out the necessity of establishing between the u and Ψ waves a bond sufficiently flexible to avoid forcing the u wave to take on any of the subjective character of the Ψ wave. We have now seen how that goal may be achieved.

Obviously, it is mathematically the same thing to write $\Psi = Cv$ or $v = C\Psi$, but the advantage of using the first form now becomes evident. It makes it possible for us to choose the constant C to suit our convenience, if need be, by attributing to it different values in separate regions of space, in such a way that the Ψ wave, *thus constructed by beginning with* v, may play the statistical role that we wish to assign it. That is why we must put $C = 0$ for the regions in space where we know that the u wave involves no singular region, since the probability of presence of the singular region is there zero.

8. The extension of the foregoing ideas to collision problems

The ideas that have just been expounded, based on the example of the semi-transparent mirror, may be generalized to the case of collision phenomena.

Let us base our argument on the case of two colliding particles. The theory usually assumed takes for the initial state two trains of Ψ waves occupying separate regions of space R_0 and R'_0 and approaching each other. Then, after the very complex phenomenon of collision, which the currently-held theory can represent only by means of six-dimensional configuration space, one finally arrives at a terminal state in which the Ψ wave-trains of the two particles are once more entirely separated and in which there are several possibilities $R_1, R_2, \ldots$, for the position of the first particle's wave-train, and several possibilities $R'_1, R'_2, \ldots$, for the position of the wave-train of the second particle. These wave-trains are "correlated", that is, if particle 1 is in R_1, particle 2 is in R'_1, if particle 1 is in R_2, particle 2 is in R'_2, etc.

In the initial state, according to the conceptions of the Double Solution, the u_0 and u'_0 waves of the two particles occupy R_0 and R'_0; the corresponding Ψ waves $\Psi_0 = C_0 v_0$ and $\Psi'_0 = C'_0 v'_0$ are merely representations of probabilities, and the constants C_0 and C'_0 must be chosen in such a way that the Ψ waves may play that role. In the final state, the u wave of the first particle will be divided and shared by $R_1, R_2, \ldots$, and that of the second particle by $R'_1, R'_2, \ldots$, but the two singular regions must always be situated in two correlated wave-trains (for example, the first in R_1 and the second in R'_1). The fictitious

Ψ waves must be constructed in such a way that the square of their modulus represents the probability of presence of the singular regions when we do not know in what two correlated wave-trains they are found. When an observation or a piece of information tells us in which correlated pair of wave-trains the particles are situated, we must then renormalize the Ψ waves in such a way that all of them are zero in the correlated pairs of wave-trains, except in the pair where the particles are found. Naturally that will in no way modify the value of the u waves in the various wave-trains, since the u wave is in an objective reality independent of our information. The Ψ waves will thus find

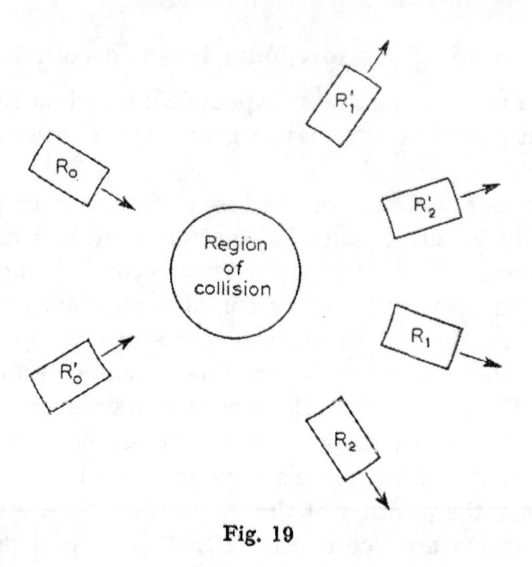

Fig. 19

themselves concentrated, by the process of renormalization, in the correlated pairs of wave-trains in which the singular regions are found, with the coefficients C of the relation $\Psi = Cv$ becoming zero in all the other correlated pairs of wave-trains. This is easily understood, for the Ψ waves, being a mental construct, are not in reality bound to the objective structure of the u wave and its local values, but must simply make it possible to represent the probability of presence of the "singular-region particles".[7]

[7] Here again one might assume that the u wave-trains containing a singular region are "regenerated", while those without singular region degenerate and perhaps disappear.

9. Summary

In this chapter we have studied three questions that are very difficult to answer from the point of view we have adopted. They are: the particle associated with a wave-group in free propagation, the emission of particles by an isotropic point-source, a particle passing through a mirror which is semi-transparent to the particle's associated wave.

We found ourselves confronted with the following dilemma: We had either to admit that the external portion of the u wave weakens progressively in such a way that the particle tends to "lose" its outer u wave, or to assume that, on the contrary, the particle keeps its initial u wave entirely intact. The first hypothesis corresponds to the form of the solutions that are habitually considered in Classical Physics and usual Wave Mechanics for the *linear* equations of continuous waves. In the theory of the Double Solution this would entail admitting that a difference must exist between "young" (*i.e.*, recently emitted) particles and "old" ones (*i.e.* those which have undergone numerous interactions since their emission and which are partially deprived of their external u wave). But in that case it would be rather difficult to understand the fact—one that is, however, almost a certainty—that the "old" particles have the same properties (especially the same interference properties) as the "young" ones. Indeed, one must not forget that interference phenomena are easily produced with photons coming from distant stars that have travelled through space for immensely long times.[8] The second hypothesis, on the other hand, seems in the theory of the Double Solution more satisfactory from the physical point of view, because it would better correspond to the stable nature of particles. But in order to justify the second hypothesis, it would be necessary to show that it corresponds to those properties of the u wave which are bound essentially to the *non-linear* character of its equation of propagation. Unfortunately, complete development of this justification—made extremely complicated from the very outset by the difficulty always present in studying the solutions of non-linear equations—will remain impossible so long as we remain in ignorance of the form of the non-linear terms to be introduced into the equation of the u wave.

[8] One might, however, seek to interpret the apparent "receding" of the spiral nebulae in terms of the "aging" of particles rather than by invoking the hypothesis of the "expanding universe".

Chapter XIX

STATIONARY STATES, QUANTUM TRANSITIONS AND THE CONSERVATION OF ENERGY

1. Stationary states

One of the most basic concepts introduced by the older quantum theory was the stationary state of a quantized system. In his theory of the atom, Bohr, in 1913 introduced the idea that an atom can be in only a certain number of quantized states to the exclusion of all others, and he connected this idea with Planck's quantum theory. The atom is said to be capable of passing directly from a stationary state of energy E_i into a different stationary state of energy $E_k < E_i$ by emitting a quantum of radiation of frequency ν_{ik} such that $h\nu_{ik} = E_i - E_k$ (Bohr's law of frequencies). When irradiated by a ray of frequency ν_{ik}, the atom is also able to pass from the stationary state of energy E_k into a stationary state of higher energy E_i by absorbing a quantum of radiant energy $h\nu_{ik}$. If the atom in state E_i is subjected to collision with an incident particle, the atom can pass from the state with energy E_i into a state with energy E_k by transferring the energy $E_i - E_k$ to the incident particle. Conversely, if the atom is in a state of energy E_k, it can pass into the state of energy E_i if the incident particle possesses enough energy for the atom to take from it the energy $E_i - E_k$. That, in its main lines, is the very simple scheme of Bohr's original theory. Experimental study of the phenomena of excitation and de-excitation by collision, and also of ionization phenomena induced by collision (wherein the atom entirely loses one of its electrons), have entirely confirmed the existence of stationary states, and the principle has become one of the fundamental concepts of Quantum Physics.

In Bohr's conceptions, there are only stationary states, and any description of sudden transitions accompanying the change of such stationary states is *a priori* excluded. According to Bohr, the atom in a stationary state is, so to speak, unaffected by time; *i.e.*, it does not evolve in time. As for the sudden transition that causes the atom to pass from one state to another—that, according to Bohr, is something it is absolutely impossible to describe by any spatio-temporal picture.

When Wave Mechanics first appeared on the scene, the notion of stationary states was interpreted as follows: The Ψ wave describing the state of a quantized system has, when the system is in a stationary state, the form of a "stationary wave", meaning that the expression of the nth stationary state is

$$\Psi_n = a_n(q)e^{\frac{i}{\hbar}E_n t} \tag{1}$$

where q represents the collection of configuration variables describing the system. To determine the stationary states, one would have to find for the wave equation of the quantized system solutions of form (1) which are finite, uniform and continuous and which satisfy the boundary conditions of the problem under consideration. These solutions constitute the "eigen functions" of the Hamiltonian operator corresponding to the "energy", and the "eigen values" E_n are the energies of the quantized stationary states of the system.

It was Schrödinger who in 1926 first calculated the energies of stationary states in that way. But it is likewise he who most severely criticized the very notion of stationary states. In a recent article[1] —in which, moreover, he has strongly insisted on the basic role that the limited dimensions of the wave-trains must play in Wave Mechanics—he has amusingly pointed out that the exclusive consideration of stationary states led to a theory "which while describing minutely the so-called "stationary" states which the atom had normally, *i.e.* in the comparatively uninteresting periods when *nothing happens*, the theory was silent about the periods of transition or 'quantum jumps'. . ." He notes that in general the state of an atom must be represented by a superposition of stationary waves of the form

$$\Psi = \sum_k C_k a_k(q)e^{\frac{i}{\hbar}E_k t} \tag{2}$$

and he concludes from this that in Wave Mechanics it must be possible to represent the quantum transitions while still respecting the "prerogatives" of the eigen frequencies E_k/h, but at the same time *completely suppressing the "prerogative" of the stationary states.* These remarks are of interest, and we will return to them.

Let us add that Schrödinger would like, in keeping with his original tendency, to suppress the notion of a particle as much as possible and to consider phenomena of quantized energy exchanged through colli-

[1] *The British Journal for the Philosophy of Science,* 3 (1952) no. 10 and no. 11.

sions as resonance phenomena. We will not follow him along that path, which appears very different from ours; instead, we are going to consider how, in the theory of the Double Solution, the conservation of energy appears when the Ψ function is a superposition of eigen functions of the Hamiltonian.

2. The conservation of energy during a collision of a particle with an atom

In order to study a specific case, let us consider the collision of a particle, whose initial energy may be considered as having an exact value E_0, with an atom whose Ψ wave function has the general form (2). Calculation by Wave Mechanics shows that after collision there will be a whole series of possibilities with various accompanying probabilities. For each one of these possibilities the particle will move away from the atom with an energy of the form $E_0 \pm (E_l - E_m)$, where $E_l - E_m$ represents one of the differences of energy-levels of the quantized atom. If we find, after the collision, that the particle has an energy $E_0 + E_l - E_m$, we would have to conclude from this that the atom has undergone the transition $l \rightarrow m$ and that it finally arrives in the stationary state of energy E_m. If in its initial state the atom had been in the stationary state of energy E_l, one could simply say that, at the time of the collision, when the transition $E_l \rightarrow E_m$ takes place, it transferred the energy to the incident particle, and conservation of energy would then have a very clear meaning. But this meaning is less clear in the hypothesis we have adopted, since in it the Ψ wave function has the initial form (2). Following the usual interpretation, we should then say that, in its initial state, the atom has "potentially" all the energies E_k with the respective probabilities $|C_k|^2$. The fact that after the collision the particle may be found to have the energy $E_0 + (E_l - E_m)$ then shows that, in the collision, it is the possibility $E = E_l$ that was "actualized" for the atom; and the atom, having given up the energy $E_l - E_m$, ultimately finds itself in the state of energy E_m. In this way we see how much more difficult it is to formulate conservation of energy here than it was in Classical Physics, because the atom's energy no longer has a well determined value in its initial state. The same would naturally hold true if the Ψ wave of the incident particle involved several spectral components, even when the atom might happen to have an initially well defined energy.

Since in general the particle moves indefinitely away from the atom after the collision, the determination of the final value of its energy may be made very far from the atom. But it is precisely that determination which, according to the current interpretation, would put the atom into the energy-state E_m! Here we have again one of the aspects of the paradox of correlated states which was brought to light by Schrödinger and explained previously in Chapter VII, section 4.

3. The double-solution point of view

In the causal interpretation we are presenting, the phase φ plays, as we know, the part of a Jacobi function extrapolated beyond the domain of Geometrical Optics, and the energy of a particle (or of a system) is, consequently, given by the formula

$$W = \frac{\partial \varphi}{\partial t}. \tag{3}$$

Here we come upon a remarkable situation. Even in the absence of an external field variable in time, the energy is not in general constant. Let us take the simple case of a particle in the absence of any external field. If the Ψ wave is a plane monochromatic wave, that is, if the particle is situated in a group of almost monochromatic waves but not in the immediate vicinity of the edge of that group, the energy of the particle will remain constant. But as soon as the Ψ becomes a superposition of plane monochromatic waves with an extended spectrum, it is easily seen by calculating the phase φ that this phase will no longer be a linear function of time and that the energy of the particle defined by (3) will no longer be constant.

This fact is tied in with the following circumstance: The quantum potential Q is then an explicit function of time. Now we know that, if $\mathscr{L}$ designates the Lagrange function of a particle (or a system), the total derivative of the energy W taken by following the motion is

$$\frac{dW}{dt} = -\frac{\partial \mathscr{L}}{\partial t}. \tag{4}$$

For there to be conservation of energy $\mathscr{L}$ must, thus, not depend explicitly on time. Now we have seen in Chapter X, section 1, that in the dynamics of the Double Solution the Lagrangian schema is valid with the definitions

$$\mathscr{L} = \tfrac{1}{2}mv^2 - F - Q; \qquad W = \tfrac{1}{2}mv^2 + F + Q \tag{5}$$

(in the non-relativistic approximation), F being the potential from which the external field is derived, and Q the quantum potential. So, even if F is zero or does not explicitly contain time, the energy will only be constant if Q does not depend explicitly on time. But in the case of a superposition of plane monochromatic waves, Q generally depends explicitly on time.

Let us take the simple case of the hydrogen atom. The electron will have a constant energy only in the stationary states. In all the states where the Ψ wave has the general form (2), the electron will in the causal theory have a very complicated trajectory defined as a function of the phase φ of the Ψ wave by the guidance formula, and its energy will not remain constant in the course of motion.

When an incident particle of well defined energy E_0 happens to hit an atom, the causal theory thus leads us to believe that, in principle, the result of the collision would be entirely predictable if we could know the initial position of the particle in the incident wave-train and the initial position of the electron in the atom. However, these positions remain of necessity unknown, since every measurement of the position modifies the initial positions; we must rest content with calculating, in agreement with the formulas of the usual interpretation, the possible results of the collision, along with their respective probabilities. Performing the calculation according to the usual method, we will find the same final result: The particle must move away from the atom with a well defined energy equal to $E_0 + (E_l - E_m)$ corresponding to a gain or loss of energy equal to one of the differences of energy-levels of the atom; and, in keeping with this, the atom will remain in a final stationary state of quantized energy E_m.

We thus see that, as in the usual interpretation, there is no longer conservation of energy, properly speaking, in our own interpretation, since the atom does not in general have a constant energy in the initial state. Whereas, in the usual interpretation, the atom in its initial state would have several "potential" values for its energy, along with their various probabilities; in the causal theory the atomic electron would have a continuously variable energy in the initial state. Neither in one case nor the other are the classical conditions for well defined initial and final values met. So the situation with regard to the conservation of energy does not seem basically any better in the current interpretation than it is in the causal theory.

4. Another instructive example of a collision between an atom and a particle

By way of an example, we are going to consider another instructive case of a collision between an atom and a particle.

Let us assume an atom initially in its ground state of minimum energy E_0. This atom possesses two excited energy-states E_1, and $E_2 > E_1$. The schematic representation of its levels is then:

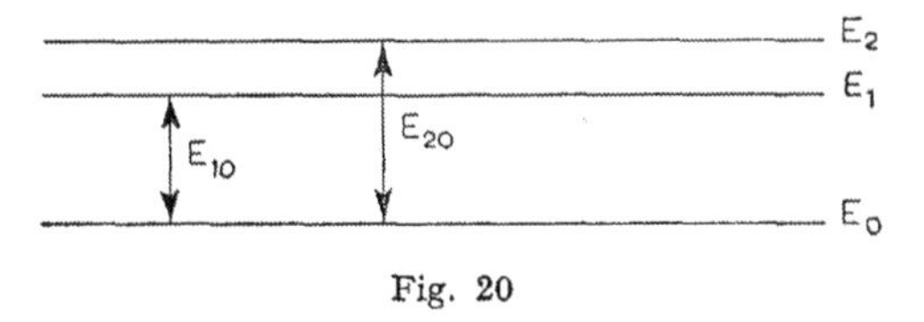

Fig. 20

Let us put $E_1 - E_0 = E_{10}$ and $E_2 - E_0 = E_{20}$: These are two excitation-energies of the atom in its normal state. One obviously has $E_{20} > E_{10}$.

We further assume that, colliding with the atom in the state E_0, there is a particle whose Ψ wave is

$$\Psi = e^{\frac{i}{\hbar}(E_{10}t - p_{10}z)} + e^{\frac{i}{\hbar}(E_{20}t - p_{20}z)}. \tag{6}$$

In reality—and this is a point that physicists generally forget to emphasize when they are concerned with these problems but which might have considerable importance—the necessarily limited Ψ wave has form (6) only in the central portion of the two overlapping wave-groups.

In the usual interpretation one has to say that the incident particle has potentially, with equal probability, the two energies E_{10} and E_{20}. Its collision with the atom may give rise in that atom to one of the transitions $E_0 \to E_1$ or $E_0 \to E_2$. In the first case the final kinetic energy of the incident electron will be zero; in the second case it will be equal to either $E_{20} - E_{10}$ or zero. One cannot say that there is conservation of energy in the classical sense, since the system does not have, in its initial state, an energy with a well defined value.

In the causal interpretation, on the other hand, the Ψ wave of the incident particle will be written in the form

$$\Psi = 2\cos\left[\frac{1}{\hbar}\left(\frac{E_{20} - E_{10}}{2}t - \frac{p_{20} - p_{10}}{2}z\right)\right] e^{\frac{i}{\hbar}\left[\frac{E_{10}+E_{20}}{2}t - \frac{p_{10}+p_{20}}{2}z\right]} \tag{7}$$

equivalent to (6), and we will have

$$\frac{\partial \varphi}{\partial t} = \frac{E_{10} + E_{20}}{2}, \qquad \frac{\partial \varphi}{\partial z} = \frac{p_{10} + p_{20}}{2}. \tag{8}$$

The energy W of the incident particle is thus equal to the constant value $(E_{10} + E_{20})/2$, which corresponds to the fact that, since the amplitude of the Ψ is

$$a = 2 \cos \left[\frac{1}{\hbar} \left(\frac{E_{20} - E_{10}}{2} t - \frac{p_{20} - p_{10}}{2} z \right) \right], \tag{9}$$

the quotient $\Box a/a$, and consequently the quantum potential Q are constants and independent of time.

Since the final energy of the incident particle is either zero or $E_{20} - E_{10} = E_2 - E_1$, we see that the variation of the energy of the system particle-plus-atom corresponds to one of the following schemes:

$$E_0 + \frac{E_{10}+E_{20}}{2} \begin{cases} \nearrow E_0+E_{10}=E_0+\dfrac{E_{10}+E_{20}}{2}-\dfrac{E_{20}-E_{10}}{2}<E_0+\dfrac{E_{10}+E_{20}}{2} \\ \searrow \left\{\begin{matrix} E_0+E_{20}= \\ E_0+E_{10}+(E_{20}-E_{10}) \end{matrix}\right\} =E_0+\dfrac{E_{10}+E_{20}}{2}+\dfrac{E_{20}-E_{10}}{2}>E_0+\dfrac{E_{10}+E_{20}}{2}. \end{cases} \tag{10}$$

Here, although we find a constant energy for the initial state of the particle, and although the initial value of the energy of the system is constant and well determined, there is nevertheless no conservation of energy, because the variation of the total energy during the collision is equal to $\pm \frac{1}{2}(E_2 - E_1)$. This variation is mathematically connected with the fact that, during interaction, there intervenes a quantum potential explicitly dependent on time. It should be noted that since the two possible transitions (10) have the same probability of $\frac{1}{2}$ and since they give rise to energy-variations that are equal but of opposite sign, there would be a kind of statistical conservation of energy similar to that considered momentarily by Bohr, Kramers and Slater after the discovery of the Compton effect.

But one may wonder if, by introducing along with the energy $\partial\varphi/\partial t$ of the particle, another energy bound to the surrounding wave, one might not be able to reëstablish conservation of energy. We will examine this question with the help of a result that I obtained in 1927.

5. The energy-momentum tensor in the pilot-wave theory

In a note in the *Comptes rendus* of November 1927 [13], I had already shown that there existed in the theory of the pilot-wave a momentum-energy tensor formed by utilizing the Ψ wave and having a conservational property. Here is the way I reasoned:

We have seen that in the causal theory a Dynamics of the particle may be developed in which the particle would have the variable rest-mass

$$M_0 = \sqrt{m_0^2 + \frac{\hbar^2}{c^2}\left(\frac{\Box a}{a}\right)}, \tag{11}$$

and we came up, using the notation of General Relativity, with the two equations

$$(\mathrm{J}) \quad g^{kl}\left(\frac{\partial\varphi}{\partial x^k} - e\mathrm{P}_k\right)\left(\frac{\partial\varphi}{\partial x^l} - e\mathrm{P}_l\right) = M_0^2 c^2$$

$$(\mathrm{C}) \quad \frac{1}{\sqrt{-g}}\frac{\partial}{\partial x^2}\sqrt{-g}\, g^{kl} a^2\left(\frac{\partial\varphi}{\partial x^k} - e\mathrm{P}_k\right) = 0$$

e being the particle's electric charge and P_k the "electromagnetic potential" four-vector. Since the four-velocity is defined by $u^l = dx^l/ds$ with $u^l u_l = 1$, the guidance formula is written

$$M_0 c u^l = g^{kl}\left(\frac{\partial\varphi}{\partial x^k} - e\mathrm{P}_k\right). \tag{12}$$

Since there arises from (C) the fact that the four-vector of covariant components equal to $a^2(\partial\varphi/\partial x^k - e\mathrm{P}_k)$ has a zero divergence, we may assume it to be proportional to the four-vector

$$\mathrm{C}^l = \rho_0 u^l \tag{13}$$

which expresses the density and the flux of the particles in a cloud of particles describing all the trajectories associated with a single Ψ function. One is then led to write

$$\rho_0 = \mathrm{K} M_0 c a^2. \tag{14}$$

The density of the cloud, since it is given by the fourth component of the vector C^l, has the value

$$\rho = \mathrm{C}^4 = \mathrm{K} M_0 c a^2 u^4 = \mathrm{K} a^2 g^{4k}\left(\frac{\partial\varphi}{\partial x^k} - e\mathrm{P}_k\right). \tag{15}$$

One can easily see that, in this Dynamics, the equations of the particle's motion take on the form

$$\frac{d}{ds}(M_0 c u_l) = \tfrac{1}{2} M_0 c u^i u^k \frac{\partial g_{ik}}{\partial x^l} + e u^i \left(\frac{\partial \varphi_i}{\partial x^l} - \frac{\partial \varphi_l}{\partial x^i} \right) + c \frac{\partial M_0}{\partial x^l}. \quad (16)$$

In the right-hand member of this equation, the first term represents the action of the gravitational field; the second represents the action of the electromagnetic field, and the third the action of the quantum potential.

Let us assume there is no gravitational field, and let us continue considering the cloud of particles associated with the same Ψ wave. Multiplying (16) by $M_0\, a^2$, and taking equation (C) into account, we obtain, after a few transformations,

$$\frac{\partial}{\partial x^k}(T_i^k + \Pi_i^k + S_i^k) = 0, \quad (17)$$

where the S_i^k are the components of the well known electromagnetic momentum-energy mixed tensor. The quantities T_i^k and Π_i^k are given by

$$T_i^k = \rho_0 M_0 u^k u_i$$
$$\Pi_i^k = K \frac{\hbar^2}{2c} g^{lk} \left[2 \frac{\partial a}{\partial x^i} \frac{\partial a}{\partial x^l} - g_{il} \left(g^{mn} \frac{\partial a}{\partial x^m} \frac{\partial a}{\partial x^n} + a \Box a \right) \right]. \quad (18)$$

The T_i^k are the mixed components of the usual particle momentum-energy tensor for a cloud of proper density ρ_0, made up of particles having a rest-mass M_0. As for the tensor Π_i^k, it represents a kind of collection of internal tensions within the cloud of particles similar to the internal tensions within a fluid. If we neglect the electromagnetic field, we see that equation (17) leads us to the following conclusion: In the absence of any electromagnetic field, the energy and momentum corresponding to the sum of the tensors T_i^k and Π_i^k are left intact.

One might be tempted to utilize this result to restore conservation of energy in the causal theory. We have seen that, except for the case of the plane monochromatic wave, the motion of the particle defined by the guidance formula corresponds to continually variable energy and momentum. Since the energy and momentum of the particle are connected with the tensor T_i^k, one might try to interpret the appearance of the tensor Π_i^k by saying that an energy and momentum exchange between the particle and the outer portion of its u wave is constantly going on, and this exchange is represented by the appearance of the quantum potential. Relation (17) would then be interpreted by saying that it expresses the conservation of *total* energy and *total* momentum of the particle and its external wave. If, in the examples studied in the foregoing sections, we have found no conservation of energy, that would be simply because we had not taken into account the energy of the regular portion of the u wave.

Unfortunately, on further reflection, this interpretation does not appear acceptable. The reasoning followed above in obtaining a momentum-energy tensor starts with the statistical Ψ wave and defines the current-density four-vector C^i and the tensor T_i^k by identifying a^2 with the average density of a fluid. Now, this has meaning only for an infinite number of particles describing all the trajectories defined by the Ψ wave, but no meaning for a single particle describing a certain *specific* trajectory. The theorem of conservation expressed by (17) has, thus, only a statistical meaning, and one does not obtain by means of it a statement of conservation applicable to a single particle, as would be required in order to have real conservation in the causal theory of the individual motion.

Obviously, one could define a tensor ϑ_i^k starting with the u wave, the way Π_i^k was defined above by starting with the Ψ wave. One would have only to adopt for ϑ_i^k expression (18) of Π_i^k in which a will have been replaced by f—a definition that would, of course, be valid only outside the singular region. This new tensor—proportional, be it noted, to Π_i^k—might define tensions in the external part of the u wave; but I do not see how, with these premises, one could possibly demonstrate that there is an exchange of energy and momentum between the singular region and the regular portion of the u wave which would assure the collective conservation of these quantities.

Thus—unless there exists some way of "salvaging" conservation of energy and momentum of which I am unaware at the moment [2]—it does seem that the causal theory must assume that, apart from a few exceptional cases, this conservation is simply statistical.

6. A reconsideration of measuring processes

Taking into account what has just been said, we are going to recon-

[2] One might perhaps assume that the particles are able to exchange energy and momentum with what we call "empty space" or Bohm-Vigier's subquantic medium.

sider the interpretation of measurements in the Double Solution. Our argument will be based on the consideration of energy, but what we have to say may be transposed to the case of any other measurable quantity; moreover, we are going to consider a particle; but once more, if the necessary changes in terminology are made, these considerations may be applied to a whole system.

As an example, let us consider the following case: In the initial state we are dealing with a particle whose external u wave (to which the fictitious statistical Ψ wave must be proportional) is in general formed by a superposition of eigen functions of the Hamiltonian operator. Every process for measuring energy must finally result in a division of the original wave-train into trains of almost monochromatic waves separated in space and corresponding to a well defined value of the energy.

If $\Psi = \sum_k c_k \Psi_k$ represents the expansion of the Ψ wave in its initial state (and consequently of the external portion of the wave, within a multiplicative factor of the v wave), the interpretation usually assumed leads us to say that in this initial state the particle does not have a well determined energy, but only possible energies E_k, each one of which has a probability $|c_k|^2$ of actually becoming a reality in a subsequent measurement of the energy. The causal theory will say, on the other hand, that in aninitial state a particle has a complicated motion and a continually variable energy. The process of measuring the energy, by dividing the initial wave-train into wave-trains separated in space and almost monochromatic in character and by obliging the particle to remain *fastened* to one of these wave-trains, makes it necessary for the particle finally to have a determinate value for its energy. Without having to introduce the division of the wave-train, one might be content to say that the measuring process unfastens the v wave's original singular region and fastens it onto one of the monochromatic components of v.

Neither of these opposing interpretations satisfies the conditions of true conservation of energy. They both imply the active interference of the measuring process—an interference which completely modifies the state of the particle, in keeping with one of the basic and seemingly indisputable ideas of Quantum Physics. In the theory of the Double Solution, the active interference effected by the measuring of the energy would be precisely *the effect of the "pinnacle" or "finger"* (*making up the singular region*) *on one of the monochromatic components of the original wave, in such a way that the particle which is finally fastened to one of the monochromatic components would then come to possess a constant energy.*[3]

Similar considerations may be developed for other examples, such as the collision of a particle and an atom, where each of the as yet separate constituents has at the outset a state represented by a superposition of the eigen functions of their respective Hamiltonians. In every case conclusions similar to those just given would be found.

It seems that the interpretation offered by the theory of the Double Solution may be quite as readily admitted as the usual probabilistic interpretation; it even has the advantage over the probabilistic interpretation of presenting to the mind a clear picture "with image and motion" and of escaping from the previously studied objections (Chapter VII) that may be brought against the usual interpretation. Von Neumann's entire theory of measurement might be taken up again from the standpoint of the causal interpretation. An interesting study remains to be done on this subject.[4]

7. Reconsideration of the question of stationary states and quantum transitions

As we have said, Schrödinger has quite properly remarked that the initial state of a quantized system is not, in general, a stationary state,

[3] Mathematically this idea of a "pinnacle" may be represented in the following manner:

Let there be a particle whose initial state corresponds to the normalized Ψ wave, $\Psi_0 = \Sigma_k c_k \varphi_k$ with $\Sigma_k |c_k|^2 = 1$, the φ_k being eigen functions of the quantity to be measured. If after the act of measurement we learn that the quantity that was to be measured has a value corresponding to the index l, we will be obliged to replace Ψ_0 by $\Psi_f = \varphi_l$. This is the form that the reduction of the probability packet—a subjective operation performed after obtaining the information—takes on here.

From the objective point of view of the u wave, we must assume that in the initial state the u wave has, outside the singular region, the form $u_0 = 1/C\,\Sigma_k c_k \varphi_k$, C being the constant with a well determined physical value in the relation $\Psi = Cv$. The point of view suggested in the text leads us to say that, during the measurement, the particle has become unfastened from the u_0 wave in order to fasten onto the component with l index. The "implanting" of the singular region onto this component would mean that an objective process of finite duration, associated with non-linearity, would cause all components other than the one of index l to vanish and would reinforce this latter one in such a way that the u wave would finally have the form $u_f = (1/C)\,\varphi_l$, C still being the same constant. After the reduction of the probability packet, the relation $\Psi = Cv$ would, thus, ultimately be restored.

[4] Let there be a measurable quantity A bound to a particle, and let φ_k be the eigen functions corresponding to A. If the initial state is a superposition of the φ_k, the theory

but a superposition of stationary states ($\Psi = \sum_k c_k \Psi_k$), and this deprives the stationary state of the unjustified prerogatives that have often been conferred upon it since Bohr originally set forth his theory. And this is just as true in the causal interpretation as it is in the usual one—with one difference: The usual interpretation holds that it is impossible to describe the state of the atom in terms of a motion in space, corresponding, so to speak, to several possible energy values having probabilities equal to the quantities $|c_k|^2$; the causal interpretation, on the other hand, considers that this same state corresponds to a motion perfectly determined in the course of time, but having a constantly varying energy.

The emission of a quantum of radiation by the atom then appears as equivalent to an act of measuring the energy in the sense that, when the emission is over, the photon emitted is associated with a wave-group that has become separated from the quantized atom, and by observing this photon one can find out in what final quantized state the atom is and, consequently, one can attribute to the atom a definite and constant energy.

The case is thus analogous to those studied in the preceding section. One can indeed, although this may be just a manner of speaking, consider the atom in its initial state as equivalent to a system made up of an atom and an annihilated photon of zero energy. We are thus brought back to the case of a system made up initially of a particle with well determined energy and an atom whose state is represented by a superposition of stationary waves. The probabilistic interpretation then tells us that in the initial state, the system has a whole series of possible E_k values for its energy, each with the probability $|c_k|^2$, and that in the final state, after a transition that it is impossible to describe by any spatio-temporal picture, the system has taken on a stationary state corresponding to one of the quantized energies E_i, with the photon carrying off an energy equal to the difference $E_j - E_i$ of the two quan-

---

of the Double Solution will hold that the singular region of the particle's u wave is originally implanted on the wave $v = \Sigma_k c_k \varphi_k$. The double-solution point of view will add that, as a result of the interactions involved in a measurement of A, the singular region will finally be implanted on a v wave proportional to one of the φ_k. In the terminology of von Neumann, we would say that there is a transition from the initial "pure case" to the final "mixed case" when the singular region, becoming unfastened from the initial superposition, has actually settled on one of its components *without our yet knowing which one.*

tized energies of the atom. From this point of view, everything would take place, then, as if the atom had first of all selected the stationary state E_j from those of the original superposition and had then undergone the transition $E_j \rightarrow E_i$, accompanied by the emission of the photon.

The point of view of the causal interpretation is, naturally, entirely different. Here the initial motion of the constituent parts of the atom may be perfectly described by a spatio-temporal image, but the initial motion will correspond to a continually variable energy. As the result of a series of successive states, still describable, in principle, by a spatio-temporal picture which would constitute the quantum transition, the system would reach a final state wherein the photon would be emitted in the form of a wave-group separated from the atom, and the atom would then find itself in a final quantized energy-state E_i, with the photon having the energy $E_j - E_i$. Here again everything finally takes place as if the atom had selected the quantized state from those of the original superposition and then transferred to the photon the energy $E_j - E_i$, with conservation of energy. But we have here passed from the initial to the final state by a well defined motion that permits a description of the quantum transition in terms of space and time. According to what we found previously, this motion would, moreover, take place without conservation of energy—such conservation taking place only statistically as an average for a very large number of similar processes corresponding to various initial values for the position of the constituents of the system.

Consider the case where the emission of the photon may be attributed to the change of state of a single intra-atomic electron—as is the case of the hydrogen atom. Then, in the initial state, the singular region that constitutes the atomic electron is fastened to a v wave corresponding to a superposition of stationary waves. The departure of the photon after a period of agitated motion will have had the effect of detaching the electron from the initial v wave in order to fasten it to one of the monochromatic components, and this will have made its energy constant. We here have an image that was employed in the previous section.

We must now emphasize strongly a point of very great import. In the present state of the theory, the probabilities of the processes of quantum transitions are evaluated by a calculation that was first suggested from the correspondence principle which represents the action of matter on the electromagnetic field by expressions in which the electromagnetic field of classical Maxwell-Lorentz wave theory occur, and in which the electrified elements of matter (intra-atomic electrons) appear via the current-density four-vector defined by starting with the Ψ wave of the atom. For almost thirty years now it has been pointed out that there is here a sort of contradiction with the manner in which the equation for the intra-atomic waves is written, for in the latter equation the Coulomb interaction potentials between charged particles are involved, which amounts to regarding them as point-like and well localized in the atom. So, on the one hand, in order to obtain the form of the Ψ wave, the electrons are considered well localized inside the atom, and, on the other hand, in order to calculate the interaction with the radiation, these same electrons are considered as if they were smeared throughout the atom with a statistical density $|\Psi|^2$. This flagrant contradiction can hardly be explained without adopting the following conclusion: The entire present theory of interactions between matter and radiation and the prediction of the transition probabilities that derive from it have only a statistical value. They allow exact prediction of the total phenomena of emission, absorption, diffusion, etc., but they furnish no precise description at all of the individual phenomena. The present purely probabilistic theory extricates itself from this difficulty by, it seems, denying the very existence of the individual phenomena. But not only does it seem to me very difficult to make such an admission; it likewise strikes me as being in contradiction to the use of the Coulomb potential for the representation of the interactions.

Still another reason that leads me to believe that the current theory of interactions between matter and radiation has only a statistical value is the fact that the classical electromagnetic field is without doubt nothing but a Ψ wave of several components, for the particle of spin 1 known as the photon. This is clearly evident from the general theory of particles with spin, especially in the "fusion theory" form that we gave to it.[5] In the theory of the Double Solution, the Ψ wave, being merely fictitious and statistical, cannot lead to a real and individual description of the phenomena, although it can, naturally, supply us with correct statistical predictions. The causal theory thus leads us to think that the true description of the electromagnetic field and its

[5] See *Théorie générale des particules à spin (méthode de fusion)*, 2nd ed., Gauthier-Villars, Paris, 1954.

interactions with the electrified particles should bring in, not the Ψ wave of the photons, *i.e.* the classical electromagnetic field, but their u wave, *i.e. an electromagnetic field with singular regions.*

It is obviously very difficult at the present time to imagine how we might describe the interaction between the u wave of the electrons and the u wave of the photons (which, being bosons, may group themselves severally along one and the same wave) in order to obtain a truly individual description of the phenomena of emission, absorption, diffusion, etc. In order to achieve this satisfactorily one would, undoubtedly, first of all, have to arrive at a representation—as I proposed in my fusion theory—of particles of spin other than $\frac{1}{2}$, as well as of the particles that could result from the fusion of particles of spin $\frac{1}{2}$, which, in the language of the Double Solution, would be represented by saying that their singular region is formed by the confluence of several singular regions merging into a single one of a generally different symmetry. The inverse phenomenon of the splitting up of a particle into several other particles ought then to be represented by a fragmenting of the singular region into several singular regions. According to presently available data on particles, they seem to be able to "change into each other practically without restriction, so long as it is compatible with the laws of conservation" [6] (in particular charge-conservation and spin) and this fact seems to indicate that it should be possible for the fragmentation in question to take place in several ways. With these conceptions, it should be possible to interpret the emission and absorption of photons, the creation and annihilation of electron pairs, and, more generally, the whole ensemble of phenomena of this kind that are now constantly cropping up in the New Physics of nucleons and mesons. By following this path, and through the formalism of second quantization, one ought to be able to find—at least as an average statistical representation—the quantum field theory with appropriate occupation numbers. One might perhaps succeed in this way in understanding the true meaning of those methods of calculation that, at the present time, look more like "recipes" for prediction than like theories that actually explain things.

These problems are most certainly very difficult, and it seems premature to tackle them in the present state of the theory of the Double Solution. But difficult does not mean impossible, and what is insoluble today may be solved tomorrow.

[6] W. Heisenberg, *Nuclear Physics*, Philosophical Library, 1953, p. 49.

Chapter XX

SUMMARY AND CONCLUSION

1. An overall view of the results obtained

If we now try to summarize the content of the second part of the present work, we may say that Chapters VIII through XVI contain an exposition of ideas already found in my publications of 1927 but here completed by subsequent research. The problems taken up in those chapters seem to us, on the whole, fairly well formulated. The transition from the Wave Mechanics of a single particle in a given field to the Wave Mechanics of interacting particle systems, however, requires filling in and needs to be completed by a detailed interpretation of Pauli's principle; in addition, the justification of the statistical role of $|\Psi|^2$ should be more rigorously worked out. Although there obviously remains much work to be done along these lines, I do not feel that any insurmountable difficulties exist in this area.

The problems touched upon in Chapters XVII, XVIII and XIX, however, are of a much more delicate sort. The relation established between the external form (or regular portion v) of the u wave and the form of the Ψ wave seems to me absolutely essential if we are able to reconcile the u-wave conception with the successes of the usual interpretation; but we have had to try to make this bond loose enough to avoid causing the u wave—which is, by hypothesis, an objective reality —to share in any way the subjective character of the statistical Ψ wave. The study of the difficulties presented by the case of wave-groups, by that of semi-transparent mirrors, and more generally the study of al those difficulties arising when the reduction of the probability packet is effected—in all of which the usually unmentioned existence of wave-fronts plays an important part—this study has led us to some interesting but daring ideas which are as yet, we must freely admit, only suggestions. The rigorous development of these ideas will necessitate difficult mathematical operations and will indeed be possible only if we succeed in arriving at a precise formulation of the non-linear equations of propagation satisfied by the u waves—perhaps through analogies with General Relativity, as Vigier hopes. Finally, the question of the conservation of energy and the whole set of considerations set forth in Chapter XIX require further painstaking investigation.

One of the basic ideas we have arrived at seems to us the following: *The usual theory, by limiting itself* a priori *to linear equations of propagation, precludes local irregularities resulting from non-linearity (such as singular regions and wave-train boundaries). In this way it obliterates particle structures and, consequently, finally achieves a continuous picture of only statistical character. The achievement of an adequate statistical image in this way is made possible by the fact that the regular portion of the u wave happens to be, in its analytical form, closely related to the form usually assumed for the Ψ wave—at least to the extent permitted by the reservations we have stated.*

2. The similarity between the conceptions of the Double Solution and much older ideas

It is curious to note that the development of the theory of the Double Solution has led us to ideas, albeit in somewhat modified form at times, that had already been suggested by various scientists during the crisis in Theoretical Physics brought on by the appearance of quanta.

From the very outset of his work on light quanta, Einstein emphasized the fact that a light wave (which, as we now know, is the Ψ wave associated with a photon) was as a kind of "ghost wave" that gave only a statistical description of the distribution of the photons. If we admit, said Einstein, the existence of an objective reality, this way of looking at things leads us to believe that photons must be some sort of singularity in the real light field, and the classical continuous light wave then furnishes only a statistical representation. Now that is precisely the idea that has served as the basis for the theory of the Double Solution.

Later Einstein, reflecting on the wave-particle dualism and on the success of the probabilistic interpretation of Wave Mechanics, was led to believe that the probability of presence $|\Psi|^2$ must result from a sort of hidden motion, of Brownian character, of the particles. We have seen that the "guidance law" leads, in general, to very complex particle movements, which cannot, however, be called Brownian. But we have also seen that the inevitable fluctuations of external potentials, as well as the fluctuations of boundary conditions which give rise to fluctuations in quantum potentials,[1] must be added to the complexity of the motions predicted by the guidance law and given them a random

[1] And perhaps also interactions with a subquantic medium.

character that permits us to call them "quasi-Brownian". The picture thus obtained seems to come very close to the one Einstein had in mind.

The conceptions introduced in Chapters XVII and XVIII may also be compared with certain observations put forward at a much earlier date. Schrödinger, at the outset of his famous researches, had hoped to achieve a picture of the particle by identifying it with a train of Ψ waves; but the constant tendency of wave-trains to spread out, because of the linear character of their propagation, made it impossible to seek in this way an identification of the particle with the totality of the Ψ wave-train. The ideas we have discussed in the last few chapters have led us to visualize the wave phenomenon centered on the particle as a sort of "cell"—in the biological sense—made up of 1) a central nucleus, which would be the singular region or particle in the strict sense of the word, 2) an extensive outer region (where $u \simeq C\Psi$), and 3) a sort of envelope formed by the wave-fronts and having non-linear properties. This kind of a cell would, in a general way, owe its autonomy and stability to the occurrence of non-linear phenomena. The totality of the "particle" entity, in the wider sense of the term, would thus be identified with a wave-train organized around a center which forms an unbroken continuity with the rest of the wave-train and which is endowed with a certain permanence. So we find once more, in a suitably modified form, Schrödinger's original idea.

We have also seen that the validity of the relation $\Psi = Cu$ in the external region of the wave-trains brought with it a kind of justification of the pilot-wave theory (in spite of the fundamentally different character of the u and Ψ waves, the one being objective, the other subjective) and, without allowing us to treat the Ψ wave as a physical reality, the validity of the relation explained the success of Bohm's point of view.

Finally, on examining the problem of the emission of particles by a source and of the representation of this phenomenon by a divergent spherical wave, we have recognized that the divergent spherical wave could only be an average representation of the isotropic emission, by the source, of wave-trains constructed in the way that we have just described. In this way we approach something like the conception of "needle-radiation" (*Nadelstrahlung*) once developed by Einstein, and according to which a point-source would send out limited wave-trains in all directions.

This collection of comparisons shows that the causal theory of the

Double Solution might serve to bring about a synthesis of various earlier attempts to represent the wave-particle dualism in a concrete manner.

3. Possibilities of experimental verification

Before the causal theory can really impose itself upon us, it would have not only to obviate the objections at present brought against the purely probabilistic interpretation (something it seems already to have done in large measure), but it would also have to lead to the prediction of perceptible phenomena which the current interpretation does not predict. Without being able to assert that such will be the case, we can nevertheless indicate along what lines such confirmation might be sought.

First of all, in the field of Nuclear Physics, where theory is still rather embryonic and rather infrequently successful, the fact that in a space with dimensions no larger than 10^{-12} cm there would be found, according to the theory of the Double Solution, a piling up of great number of singular regions, leads us to surmise that the usual interpretation might here be inadequate. In fact, from the point of view of the theory of the Double Solution, the justification of the usual interpretation of the role of the Ψ waves is based upon the hypothesis that the singular regions are equidistant from each other, and the distances would be large in proportion to the dimensions of these regions; that is, they must in no way overlap. It is quite possible that this condition will be found not always to prevail in nuclei, and then the predictions based upon the statistical properties of the Ψ wave would prove inadequate. A new sort of description of the nucleus utilizing u waves with singular regions very close together or even overlapping might permit the correct forecasting of certain nuclear phenomena.

We have pointed out (Chapter XI, section 6) that the demonstration of the guidance formula implies the hypothesis that the common phase of the u and Ψ waves has the same value over the entire sphere S, with which we surrounded the singular region. Now this hypothesis would of necessity cease to be correct for particles having sufficiently high energies, and then the statistical significance of the Ψ, which in the theory of the Double Solution is derived from the guidance formula, might no longer be valid. Since advances in experimental techniques make it possible to obtain particles of increasingly greater energies, one may conjecture that the moment will arrive when predictions made by utilizing the Ψ function in the way the current interpretation does will cease to be valid. It would then be necessary to try to find out what happens in the propagation of u waves when their wave-length reaches the order of dimensions of the singular region—a project that should permit prediction of observable phenomena beyond the limit of validity of the use of Ψ waves.

If there exist, in accord with the concepts put forth in Chapter XVIII, unspreading wave-groups, it could also be that some of the properties usually attributed to wave-groups by the linear theory are not correct and that from this circumstance result certain observable facts that do not conform to the usual predictions—such as, for example, a limitation of the narrowness of the beams that may be isolated by taking a very small spectral interval out of a continuous band.

Another avenue of verification for the theory of the Double Solution ought to open in the area of particles on the atomic scale. We know that this theory has, at the present time, run into serious difficulties, especially as regards infinite self-energies. A description of elementary particles that would identify them with a singular region of the u field in a spatio-temporal framework would permit us to go back to the notion of a "radius" of the particle (an electron radius, for example) and thus to avoid the pitfall of infinite self-energies. One might also—perhaps by introducing the idea of fusion—reduce the properties of particles, such as spin, magnetic moment, or even of mass, to structural differences corresponding to differences in the form of the u wave inside the singular region. If these hopes were fulfilled, one might achieve, along these lines, a description and a natural classification of particles which the constant discovery of new sorts of mesons makes more desirable every day. Granted that all this is merely a study-plan made extremely difficult to put into effect, no matter how one looks at it, by our ignorance of exactly what goes on inside the singular regions. Nevertheless, it is not forbidden to hope that the causal theory, by allowing us to represent the properties of the particles in a spatio-temporal framework, may one day supply a genuinely clear and explanatory theory of the properties of particles. By contrast, that hope seems utterly beyond realization in the framework of the present probabilistic interpretation, for the probabilistic approach has as its sole instruments of description a Ψ wave of statistical and subjective character, and abstract formalisms—likewise statistical—such as those of second quantization and the quantum theory of fields. So, within the framework of some future particle theory, experimental verifications of the theory of the Double Solution appear to be possible.

4. The agreement of the theory of the Double Solution with General Relativity [2]

The similarity of the guidance formula to the demonstrations of Georges Darmois and Einstein in General Relativity leads one to think that there exists an intimate relationship between the two points of view. Vigier has pursued this line of thought with great zeal by trying to introduce u-wave functions into the framework of an appropriately defined space-time. I will not say anything about the value of Vigier's efforts, which could doubtless be modified in various ways. But it is certain that attempts of this sort hold great interest, for they might lead to the unification of the ideas of General Relativity with those of quanta.

The goal to be achieved would be to represent every type of particle (*including the photon*) as a singular region in a u-wave field properly incorporated into the structure of space-time. In this representation Planck's constant should appear in such a way as to enlighten us concerning the true meaning of the quantum of action. The very way in which the u wave of each type of particle would be defined might lead to the discovery of the form of the non-linear equations satisfied by that u wave (or by its components, when it has several of them). In this way one would obtain one of the basic facts essential to the complete developments of the theory of the Double Solution in the form we have adopted for it.

This method of defining particles as small regions wherein a certain field obeying non-linear partial differential equations has very large values is entirely in agreement with the conceptions that Einstein has always held on this subject. In fact, he has written: "What appears certain to me, however, is that, in the foundations of any consistent field theory, the particle concept must not appear in addition to the field concept. The whole theory must be based solely on partial differential equations and their singularity-free solutions." And then further on: ". . . if a field theory results in a representation of corpuscles free

[2] On this question Vigier's doctoral dissertation may be referred to.

of singularities, then the behavior of these corpuscles in time is determined solely by the differential equations of the field."[3] Let us emphasize that, in the region we call singular, the *u* function must have very large values, but probably not a true mathematical singularity. And that is in agreement with Einstein's conceptions. The guidance theorem, in fact, agrees perfectly with the sentence just quoted from Einstein.

Einstein has called these fields containing strong local condensations, which he thinks must be the true representation of particles, "bunch-like fields". In our conception the *u* waves are indeed bunch-like wave-fields.

Fulfilling a hope expressed on repeated occasions by the supremely gifted physicist who, in the course of a single year, discovered Relativity and light quanta, the *u*-wave theory may perhaps one day help to achieve a magnificent synthesis of General Relativity and Quantum Theory.

[3] Albert Einstein, *Ideas and Opinions*, Crown Publishers, Inc., New York, 1954, pp. 306–307 and 320.

Appendix

AN ALTERNATIVE DEMONSTRATION OF THE GUIDANCE FORMULA

A recent work by Gérard Petiau [14] suggested to us a new demonstration of the guidance formula based on the theory of linear partial differential equations of the first order [15].

Starting with the Klein-Gordon equation, as we did in Chapter IX, we obtain as the continuity equation of the regular solution $ae^{\frac{i}{\hbar}\varphi}$,

$$\frac{\partial a}{\partial t}+A(x,y,z,t)\frac{\partial a}{\partial x}+B(x,y,z,t)\frac{\partial a}{\partial x}+C(x,y,z,t)\frac{\partial a}{\partial z}+D(x,y,z,t)a=0, \tag{1}$$

with

$$A=-c^2\frac{\dfrac{\partial\varphi}{\partial x}+\dfrac{e}{c}A_x}{\dfrac{\partial\varphi}{\partial t}-eV}; \qquad B=-c^2\frac{\dfrac{\partial\varphi}{\partial y}+\dfrac{e}{c}A_y}{\dfrac{\partial\varphi}{\partial t}-eV};$$

$$C=-c^2\frac{\dfrac{\partial\varphi}{\partial z}+\dfrac{e}{c}A_z}{\dfrac{\partial\varphi}{\partial t}-eV}; \qquad D=\frac{c^2}{2}\frac{\Box\varphi}{\dfrac{\partial\varphi}{\partial t}-eV}. \tag{2}$$

If the phase φ is known, then A, B, C and D are determined.

For the singular solution $u=fe^{\frac{i}{\hbar}\varphi}$ of the same phase φ, we obtain the same continuity equation (1), but with the amplitude f which has a singularity replacing the continuous amplitude a.

The differential equations corresponding to the partial differential equation (1) are

$$dt=\frac{dx}{A}=\frac{dy}{B}=\frac{dz}{C}=-\frac{da}{Da}. \tag{3}$$

The first three equations (3) have three primary integrals

$$f_1(x,y,z,t)=\lambda,\quad f_2(x,y,z,t)=\mu,\quad f_3(x,y,z,t)=\nu \tag{4}$$

which, for the constant values λ, μ, ν define lines of current in space-time corresponding, as one easily sees, to the guidance formula [in the general form (36) of Chapter IX].

But we should, in addition, consider the relation $da/a = -D(x, y, z, t)dt$. Now the variables x, y, z may be expressed, with the help of (4), as functions of λ, μ, ν and t, so that $D(x, y, z, t) = F(\lambda, \mu, \nu, t)$. Since λ, μ, ν are constant along a single line of current, we have

$$\log a = -\int^{t} dt\, F(\lambda, \mu, \nu, t) \quad \text{or} \quad a = \alpha e^{-\int^{t} dt\, F(\lambda, \mu, \nu, t)} \tag{5}$$

α being a constant and the integration over t being possible due to the constant values of λ, μ, ν. The same expression (5) is valid for f, since the phase φ is the same for the regular and the singular solutions.

The theory of linear partial differential equations of the first order tells us that the general solution of (1) is obtained by writing $\alpha = \Phi(\lambda, \mu, \nu)$ where Φ is an arbitrary function. We will then have as a form common to both a and f

$$\left.\begin{matrix} a \\ f \end{matrix}\right\} = e^{-\int^{t} dt\, F(\lambda, \mu, \nu, t)}\, \Phi(\lambda, \mu, \nu), \tag{6}$$

the first factor of the right-hand side being the same for a and f, but the function Φ being different in the two cases. Since a is by definition a regular function, the first factor cannot present any singularity, otherwise there would be no regular solution corresponding to the form adopted for φ, and that would be contrary to the hypothesis. Consequently, f can present a singularity at the point x_0, y_0, z_0 in space at the time t_0 only if the corresponding Φ function presents a singularity for the values

$$\lambda_0 = f_1(x_0, y_0, z_0, t_0); \quad \mu_0 = f_2(x_0, y_0, z_0, t_0); \quad \nu_0 = f_3(x_0, y_0, z_0, t_0)$$

of λ, μ, ν. But then the function f in space-time presents a singular line defined by $\lambda = \lambda_0$, $\mu = \mu_0$, $\nu = \nu_0$; whence the theorem:

If there exists a solution u of the linear wave-equation—a solution having a singularity and having the same phase φ as a regular solution v of the same equation, then the singular point of u moves in the course of time with the motion predicted by the guidance formula.

The demonstration given above in the case of the Klein-Gordon wave equation may be presented in a more general form applicable to all the wave equations that are found in Wave Mechanics at the present time, and especially to the Dirac equation. This general form of the demonstration has the advantage of bringing out more clearly the true nature of the result obtained.

All the wave equations of Wave Mechanics allow us to obtain a hydrodynamical picture by defining a density ρ and a current density $\rho\mathbf{v}$ that is expressed bilinearly in terms of the wave function and its conjugate, both densities obeying the continuity equation

$$\frac{\partial \rho}{\partial t} + \operatorname{div}(\rho\mathbf{v}) = 0. \tag{7}$$

Let us assume that the wave equation considered allows two "coupled" solutions—the one Ψ, regular, and the other u with a mobile point-singularity, and let us assume that these two solutions have the same lines of current defined by the same vector field $\mathbf{v}$. For the solution Ψ, the density $\rho(\Psi)$ is regular; for the solution u the density $\rho(u)$ presents a point-singularity. One may write simultaneously for both $\rho(\Psi)$ and $\rho(u)$

$$\frac{\partial \rho}{\partial t} + v_x\frac{\partial \rho}{\partial x} + v_y\frac{\partial \rho}{\partial y} + v_z\frac{\partial \rho}{\partial z} + \rho \operatorname{div}\mathbf{v} = 0. \tag{8}$$

To this linear partial differential equation of the first order in ρ there correspond the differential equations

$$\mathrm{d}t = \frac{\mathrm{d}x}{v_x} = \frac{\mathrm{d}y}{v_y} = \frac{\mathrm{d}z}{v_z} = -\frac{\mathrm{d}\rho}{\rho \operatorname{div}\mathbf{v}} \tag{9}$$

The first three equations have the integrals:

$$f_1(x, y, z, t) = \lambda, \quad f_2(x, y, z, t) = \mu, \quad f_3(x, y, z, t) = \nu \tag{10}$$

which, for constant values of λ, μ, ν define a line of current in space-time. Let us put $\operatorname{div}\mathbf{v} = F(\lambda, \mu, \nu, t)$; we then find, as above

$$\begin{aligned} \rho(\Psi) &= e^{-\int^t \mathrm{d}t\, F(\lambda,\mu,\nu,t)}\, \Phi_1(\lambda, \mu, \nu), \\ \rho(u) &= e^{-\int^t \mathrm{d}t\, F(\lambda,\mu,\nu,t)}\, \Phi_2(\lambda, \mu, \nu), \end{aligned} \tag{11}$$

the functions Φ_1 and Φ_2 not being determined by the integration. Since the first factor is the same in the two expressions (11), it must be regular, and, consequently, the function Φ_2 must at the initial instant present a singularity for the values $\lambda = \lambda_0, \mu = \mu_0, \nu = \nu_0$. As previously, there results from this the fact that the point-singularity of u must follow one of the lines of current.

Applied to the Dirac equation, the result shows clearly that the guidance must in this case be defined in terms of the current four-vector, as we did in Chapter XVI.

So we have arrived at the following general statement: *If a wave equation of Wave Mechanics allows two solutions, one regular and the other with a point-singularity, both having the same lines of current, the singularity must follow in the course of time one of the lines of current.*

Let us note that the preceding demonstration still holds true if the u wave, instead of presenting a true point-singularity, has a very small singular region where it takes on very large values. In fact, Φ_2 must then take on very high values for λ, μ, ν very close to λ_0, μ_0, ν_0, and the motion of the singular region must be represented by an extremely narrow world-tube whose axis coincides with a line of current.

BIBLIOGRAPHY

[1] LOUIS DE BROGLIE, *Compt. rend.*, 183 (1926) 447; 184 (1927) 273; 185 (1927) 380; and *J. phys. radium*, 6e série, 8 (1927) 225.

[2] LOUIS DE BROGLIE, *Electrons et photons, Rapport au Ve Conseil Physique Solvay*, Gauthier-Villars, Paris, 1930, p. 115; and *An Introduction to the Study of Wave Mechanics*, Methuen and Co., Ltd., London, 1930.

[3] DAVID BOHM, *Phys. Rev.*, 85 (1952) 166 and 180.

[4] LOUIS DE BROGLIE, *Compt. rend.*, 233 (1951) 641.

[5] TAKEHITO TAKABAYASI, *Progr. Theoret. Phys.* 8, (1952) 143; 9 (1953) 187.

[6] LOUIS DE BROGLIE, *Compt. rend.*, 234 (1952) 265.

[7] LOUIS DE BROGLIE, *Compt. rend.*, 235 (1952) 1345 and 1453.

[8] DAVID BOHM, *Phys. Rev.*, 89 (1953) 1458.

[9] LOUIS DE BROGLIE, *Compt. rend.*, 235 (1952) 557; and JEAN-PIERRE VIGIER, *ibid.* p. 1107.

[10] LOUIS DE BROGLIE, *Compt. rend.*, 236 (1953) 1453.

[11] LOUIS DE BROGLIE, *La Physique quantique restera-t-elle indéterministe?* Suivi d'une contribution de JEAN-PIERRE VIGIER, Gauthier-Villars, Paris, 1953.

[12] LOUIS DE BROGLIE, *Compt. rend.*, 237 (1953) 441.

[13] LOUIS DE BROGLIE, *Compt. rend.*, 185 (1927) 1118.

[14] GERARD PETIAU, *Compt. rend.*, 239 (1954) 344.

[15] LOUIS DE BROGLIE, *Compt. rend.*, 239 (1954) 737; 241 (1955) 345.

[16] DAVID BOHM and JEAN-PIERRE VIGIER, *Phys. Rev.*, 96 (1954) 208.

[17] LOUIS DE BROGLIE, *La Théorie de la mesure en Mécanique ondulatoire (Interprétation usuelle et interprétation causale)*, Gauthier-Villars, Paris, 1957.

See also: LOUIS DE BROGLIE, *Nouvelles perspectives en Microphysique*, 115-237 Albin Michel, Paris, 1956.

AUTHOR INDEX

SUBJECT INDEX

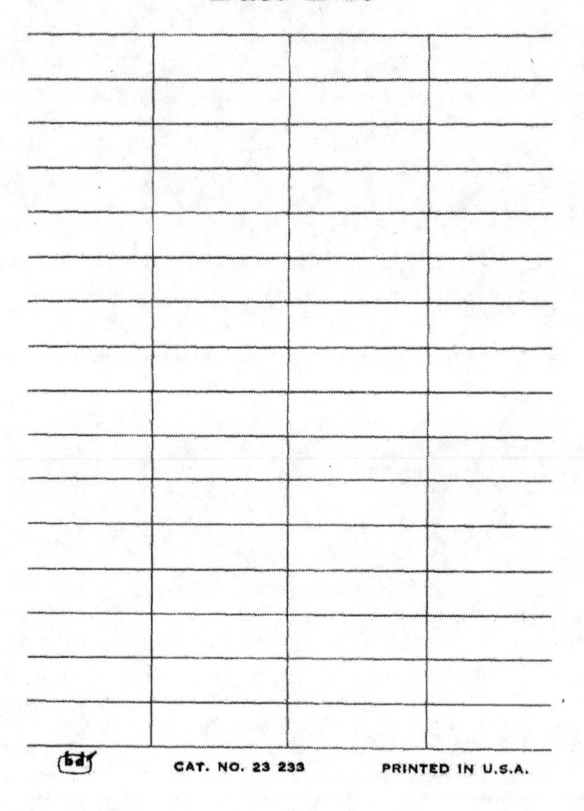
Date Due
bdy
CAT. NO. 23 233
PRINTED IN U.S.A.

www.ingramcontent.com/pod-product-compliance
Lightning Source LLC
LaVergne TN
LVHW021446300325
807252LV00003B/136

* 9 7 8 1 0 1 3 4 8 5 7 1 8 *